Cross-continental Food Chains

We live in a world of global food. The daily meals of people in both the developed and developing worlds are being transformed by the increasing ease by which food is being traded across continents. Affluent consumers' supermarket trolleys are being filled with an array of food products from developing countries while, at the same time, food exports from the developed world are supplanting and transforming dietary systems in developing countries. Some experts suggest that the enhanced tradability of food ushers in an era of increasing choice and affluence. Others point to problems of dependency, inequality and social dislocation accompanying these developments.

Cross-continental Food Chains represents a collective effort to document and understand these issues. Containing the contributions of 21 leading international social scientists from 10 countries, the book presents recent case study research on how and why the food system is being globalized, and what this means for people and communities in different parts of the world. The book covers debates on new structures and dynamics in the global trade with food products, including detailed accounts of fresh horticulture, tropical crops and livestock.

This book fills a major gap in contemporary scholarship on food and globalization. Its emphasis on case study accounts of the connections between trade and restructuring provides texture and context to these complex and important debates. Written and researched at a time in which national governments are seeking to negotiate new rules of global agricultural trade, this book is timely and relevant. It will interest researchers in geography, development studies, agricultural economics and political science, as well as professionals in the fields of trade and food policy.

Niels Fold is Associate Professor in Development Geography at the University of Copenhagen.

Bill Pritchard is Senior Lecturer in Economic Geography at the University of Sydney.

Routledge Studies in Human Geography

This series provides a forum for innovative, vibrant, and critical debate within Human Geography. Titles will reflect the wealth of research which is taking place in this diverse and ever-expanding field.

Contributions will be drawn from the main sub-disciplines and from innovative areas of work which have no particular sub-disciplinary allegiances.

Cross-continental Food Chains

Edited by
Niels Fold and
Bill Pritchard

Routledge
Taylor & Francis Group

LONDON AND NEW YORK

First published 2004
by Taylor & Francis
2 Park Square, Milton Park, Abingdon, Oxfordshire OX14 4RN

Simultaneously published in the USA and Canada
by Taylor & Francis
711 Third Avenue, New York, NY 10017

Taylor & Francis is an imprint of the Taylor & Francis Group

First issued in paperback 2011

© 2004 Wang Bin Bing

Typeset in Sabon by
Newgen Imaging Systems (P) Ltd, Chennai, India

Every effort has been made to ensure that the advice and
information in this book is true and accurate at the time of
going to press. However, neither the publisher nor the authors
can accept any legal responsibility or liability for any errors or
omissions that may be made. In the case of drug administration,
any medical procedure or the use of technical equipment
mentioned within this book, you are strongly advised
to consult the manufacturer's guidelines.

British Library Cataloguing in Publication Data
A catalogue record for this book is available from the British Library

Library of Congress Cataloging in Publication Data
A catalog record for this book has been requested

ISBN 978-0-415-33793-9 (hbk)
ISBN 978-0-415-51402-6 (pbk)

Contents

Figures

Tables

Contributors

David Barling is Senior Lecturer in Food Policy, the Centre for Food Policy, City University, London, UK.

Mónica Bendini is Professor and Director of Postgraduate Studies on the Sociology of Agriculture in Latin America at Universidad Nacional del Comahue, Argentina.

David Burch is Professor, School of Science, Griffith University, Australia.

Michelle Chauvet is Professor, Sociology Department, Metropolitan Autonomous University–Azcapotzalco, Mexico City.

Jane Dixon is Fellow, National Centre for Epidemiology and Population Health, Australian National University, Australia.

Robert Fagan is Professor of Human Geography, Macquarie University, Australia.

Niels Fold is Associate Professor in Development Geography, Institute of Geography, University of Copenhagen, Denmark.

William H. Friedland is Emeritus Professor, College XIII, University of California at Santa Cruz.

Jörg Gertel is Professor of Geography, Institute of Oriental Studies, Leipzig University, Germany.

Alex Hughes is Lecturer in Geography in the School of Geography, Politics and Sociology, the University of Newcastle, UK.

Christina Jamieson is Lecturer in Social Sciences, the Open Polytechnic of New Zealand.

Bridget Kenny is Lecturer in Sociology, the University of the Witwatersrand, South Africa.

Tim Lang is Professor of Food Policy, the Centre for Food Policy, City University, London, UK.

Richard Le Heron is Professor of Geography, the School of Geography and Environmental Science, the University of Auckland, New Zealand.

Stewart Lockie is Associate Professor in Sociology, the School of Psychology & Sociology, Central Queensland University, Australia.

Yolanda Massieu is Professor, Sociology Department, Metropolitan Autonomous University–Azcapotzalco, Mexico City.

Charles Mather is Senior Lecturer in the School of Geography, Archaeology and Environmental Studies, the University of the Witwatersrand, South Africa.

Jeffrey Neilson is a Post-Doctoral Fellow, the School of Geosciences, the University of Sydney, Australia.

Bill Pritchard is Senior Lecturer in Economic Geography, the School of Geosciences, the University of Sydney, Australia.

Norma Steimbreger is Researcher at the Grupo de Estudios Sociales Agrarios, Universidad Nacional del Comahue, Argentina.

Sietze Vellema is Programme Manager, Market Arrangements and Innovation Management, at the Institute Agrotechnology and Food Innovations, Wageningen University and Research Centre, The Netherlands.

Acknowledgements

The production of this book has been a cross-continental exercise in itself. With one of the editors in Copenhagen, Denmark, and the other in Sydney, Australia, it has involved considerable 'action at a distance'. Yet consistent with many cross-continental supply chains, the finalization of this book also demanded localized interaction, in this case, manifested through editorial meetings in Copenhagen (October 2003) and Sydney (June 2004).

In addition to the acknowledgements recognized in the individual chapters of this book, we wish to thank the Danish Social Science Research Council and the Department of Geography, University of Copenhagen, for providing financial support for the mini-conference at which the chapters of this book were initially presented.

All the chapters of this book were peer-reviewed. We wish to thank the referees for their cooperation and assistance, though for professional reasons they must remain anonymous.

Finally, we wish to thank our partners and families for their support and good humour.

Niels Fold (Copenhagen)
Bill Pritchard (Sydney)

Abbreviations

ABGC	Australian Banana Growers' Council
ABRAC	Agricultural Biotechnology Research Advisory Committee
ACP countries	Africa, the Caribbean and the Pacific
ADM	Archer Daniels Midland
AEKI	Association of Indonesian Coffee Exporters
AJCA	All Japan Coffee Association
AMH	Australian Meat Holdings Pty Ltd
AoA	WTO Agreement on Agriculture
BPS	Badan Pusat Statistik
BSE	bovine spongiform encephalopathy (mad cow disease)
CAC	Codex Alimentarius Commission
CalRAB	California Raisin Advisory Board
CAOBISCO	Association of the Chocolate, Biscuit & Confectionery Industries of the EU
CAP	Common Agricultural Policy
CBSP	Chiquita Brands South Pacific Ltd
CEC	Commission for Environmental Cooperation
CEC	Commission of the European Communities
CEN	European Committee for Standardization
CFC	Common Fund for Commodities
CGTFL	California Grape and Tree Fruit League
CI	Cook Islands
CINVESTAV	Centro de Investigación y Estudios Avanzados
CITIC	Chinese International Trust and Investment Corporation
CP Group	Charoen Pokphand Group
CTGA	California Tomato Growers Association
CTGC	California Table Grape Commission
CSO	civil society organization
CSR	corporate social responsibility
DFAT	Department of Foreign Affairs and Trade (Australia)
DFID	Department for International Development
DG SANCO	Directorate General for Health and Consumer Affairs in the European Union

EEP	Export Enhancement Program
EFSA	European Food Safety Authority
ETI	Ethical Trading Initiative
EU	European Union
EUREP	Euro-Retailer Produce Working Group
FAO	Food and Agriculture Organization
FAS	Foreign Agricultural Service of the US Department of Agriculture
FCOJ	frozen concentrated orange juice
FoRST	Foundation for Research, Science and Technology
FVO	Food and Veterinary Office
GAP	Good Agricultural Practice (EUREP)
GATT	General Agreement on Tariffs and Trade
GCC	global commodity chain
GCFG	Grampian Country Food Group
GDP	gross domestic product
GM	genetically modified; genetic modification
GMO	genetically modified organism
GRAIN	Genetic Resources Action International
GSM-102	General Sales Manager Program
HVC	high-value crop
HYV	high-yielding variety
ICO	International Coffee Organization
IITA	International Institute of Tropical Agriculture
ILO	International Labour Organization
IMF	International Monetary Fund
INP	Institute of National Planning (Egypt)
IPDL	Industrial Property Digital Library
IPPC	International Plant Protection Convention
ISO	International Organization for Standardization
ITC	International Trade Centre
IUF	International Union of Food, Agriculture, Hotel, Restaurants, Catering Tobacco and Allied Workers Associations
MOs	marketing orders
MTADP	Medium-Term Agricultural Development Plan
NAFTA	North American Free Trade Agreement
NAO	New Agricultural Country
NGO	non-governmental organization
NSW	New South Wales
OECD	Organisation for Economic Co-operation and Development
OIE	Office International des Epizooites
OJL	*Official Journal of the European Communities*
PL-480	Agricultural Trade Development and Assistance Act of 1954

RAC	Raisin Administrative Committee
RAP	US-funded Asia Regional Agribusiness Project
RBA	Raisin Bargaining Association
SAGAR	Ministry of Agriculture (Mexico)
SCAA	Specialty Coffee Association of America
SEDEX	Supplier Electronic Data Exchange
SPS	Sanitary and Phytosanitary agreement
STCP	Sustainable Tree Crop Program
TBT	Technical Barriers to Trade agreement
TESS	Trademark Electronic Search System
TNC	transnational corporation
TRIPS	Trade Related Aspects of Intellectual Property Rights
UFW	United Farm Workers union
UHT	ultra-high temperature
UN	United Nations
UNAM	Universidad Nacional Autónoma de México
UNCTAD	United Nations Conference on Trade and Development
UNORCA	Unión Nacional de Organizaciones Regionales Campesinas Autónomas
USDA	US Department of Agriculture
USDA–FAS	United States Department of Agriculture–Foreign Agricultural Service
USAID	US Agency for International Development
VAT	value-added tax
WAB	Wine Advisory Board
WGA	Western Growers Association
WHO	World Health Organization
WTO	World Trade Organization

1 Introduction

Niels Fold and Bill Pritchard

Introduction

Food has been crossing continents for centuries. The 'silk road' linking China with Europe provided a transit route for spices. The Columbian exchange introduced tobacco, tomatoes, potatoes, corn and turkeys to Europe; and transported cotton, grains, livestock, sugar and slaves to the New World. Coffee originated in Ethiopia before being introduced to the Arabian peninsula and thence to Europe (for consumption) and South America, South-East Asia and Africa (again) for propagation. British industrialization during the seventeenth and eighteenth centuries depended on working classes in the homeland being furnished with cheap grains, starches, sweeteners and meats produced in the colonies, or in areas where British capital financed an expansion of the agricultural frontier.

Recent years, however, have witnessed the cross-continental flow of foods accelerating and intensifying. For affluent consumers, supermarket aisles increasingly contain a veritable galaxy of food products sourced from across the globe. For people living in developing countries, Western food cultures, institutions and technologies (such as supermarkets, fast food chains, microwaves and refrigerators) are becoming increasingly central elements of local food systems. These developments are constituent elements of globalization. The foodscapes around us are testaments to the global political relations of the current age. By examining their detail, we observe the political contests and struggles that are integral to shaping the world in which we live.

In this book, researchers from different social science disciplines and from different parts of the world investigate the broad shape, meaning and implications of these issues. Through four major sections, richly diverse case studies address the interplay of processes upon which cross-continental food chains are hinged. The diversity of issues and themes addressed underscores the complexity of this topic area. The movement of foods from sites of production to sites of consumption animates and impacts upon a vast array of social actors, with implications for economies, cultures, dietary systems, public health and the environment. Taken together, the chapters herein seek to provide a compelling narrative of the multi-dimensional contests and struggles in the construction and restructuring of cross-continental food chains, thereby bringing into focus the political choices implicit within current structures. This perspective provides the unifying theme of this book. In a world where policy-makers often seek abstracted and simplified analysis, *Cross-continental Food Chains* asserts the need for multi-disciplinary and historically sensitized understandings of the social relations of food.

Positioning cross-continental food chains in time

The case studies reported in this book refer to the global food system of the early twenty-first century. But what are the defining features of this period, and what are its differences and similarities from previous eras?

Scholarly attention to these general questions first surfaced in the 1980s through the *food regime* concept. Not coincidently, this concept arrived in the midst of global debates on the future of agriculture, in the context of the Uruguay Round of the General Agreement on Tariffs and Trade (GATT). Food regimes are defined as extended periods during which a hegemonic political order underwrites and/or intersects with a particular system of agri-food production and trade. The seminal formulation of the concept (Friedmann and McMichael 1989) posited that between the 1870s and the 1980s global agri-food restructuring could be defined through the existence of a series of food regimes and crisis-ridden interregnums. The first food regime, which Friedmann and McMichael date from 1870 to 1914, was characterized by British Imperial hegemony. Increased international agricultural trade was predicated on the penetration of grains and livestock complexes into temperate settler-state regions (North America, Argentina, Australia, New Zealand and southern Africa) and the expansion of plantation agriculture (palm oil, cocoa) in tropical colonial territories: effectively, the relocation of the British (and to a lesser extent, European) food supply to colonial territories.

In the decades after the end of the First World War, the indebtedness of the colonial powers and the economic instability associated with the Great Depression resulted in this intersection of politics and trade losing its momentum as a driving force of global capital accumulation in agriculture. Ultimately, in the post-1945 period, a new ('second') food regime was established around the emergence of the United States as the world's largest agri-exporter. Domestically, the system of production supports and price stability furnished as part of President Roosevelt's New Deal underwrote an expansion of the soybean–hogs and grains–livestock complexes of Mid-Western and Prairies agriculture. Internationally, the hallmark of this system was the coincidence of this production regime with US political hegemony. American global political influence was buttressed by the production of large agricultural surpluses that were incorporated into the Marshall Plan and disbursed globally through food aid programmes (see Gertel this volume). The decision by the US Congress in 1947 not to ratify the proposed international trade organization kept agriculture out of the multilateral trading system, and hence made domestic US farm policies largely immune from the pressures of international trade liberalization which, by the 1970s and 1980s, impacted heavily on the manufacturing sector.

Key dynamics in this system began to alter in the mid-1980s. The institutionalization of the European Common Agricultural Policy (CAP) generated significant agricultural surpluses in Europe, encouraging trans-Atlantic agricultural trade wars. Nominal rates of agricultural protection in the Organisation for Economic Co-operation and Development (OECD) area rose from 40 per cent in 1979–81 to 68 per cent in 1986–88 (Roberts *et al.* 2001: 1). These increases placed profound pressures on national expenditure, at a time when (particularly in the US), the budgetary position of the public sector deteriorated sharply. The instability created by these conditions – including the dumping of product in developing country markets and the levying of prohibitive tariffs at short notice by both the US and EU against one another – encouraged efforts to create a new political order for world agriculture. These were given a forum in 1986, when at Punta del Este, in Uruguay, the countries of the GATT agreed to commence a new round of trade negotiations (the Uruguay Round) with the explicit objective of incorporating agriculture into the multilateral trade system and, by extension, curbing protectionism.

These initiatives have given rise to an emergent *global* regulatory architecture for governing the politics of food. The conclusion of the Uruguay Round in 1994 implemented the establishment of the World Trade Organization (WTO) as a permanent institution to promote international accord on the rules of trade, to encourage trade liberalization, and to arbitrate trade disputes. Through this, national regulation of food and agriculture has been subsumed to global institutional parameters.

Yet notwithstanding the WTO's efforts, the post-1994 arrangements represent a far cry from neo-liberal conceptions of a global 'free market' in agriculture and food. On the one hand, in combination with structural adjustment policies implemented by international lending agencies during the 1990s, the WTO framework has encouraged developing countries to restructure their agricultural sectors extensively in line with agendas to open domestic economies, liberalize land laws (thus facilitating large-scale agriculture), and to ease restrictions on foreign investment. The story is far different, however, in Northern countries. The WTO Agreement on Agriculture (AoA) broadly sanctioned the continuation of agricultural support policies in Europe, North America and Japan, albeit in restructured formats to accommodate WTO provisions. Recent developments have further emphasized the discontinuities between the neo-liberal ideal of 'free market agriculture' and the practice of agricultural policy by Northern countries. In 2002, ratification of the US Freedom to Farm Act confirmed US$98.5 billion over the following ten years for the continuation of current agricultural support programmes, and to this was added a further US$73.5 billion in new programmes. A few months later, the EU's Mid-term Review of the CAP forecast a rise in annual agricultural payments from €37.7 billion in 2001 to €41.8 billion in 2006 (Commission of the European Communities 2002: 34). This has encouraged a situation where, to paraphrase one analyst, comparative advantage becomes comparative access to subsidies (McMichael 1998: 97). Furthermore, despite the avowed purpose of the WTO to implement *trade* agreements, much of its activity has concerned issues that are not strictly about trade, such as the protection of corporate intellectual property and the facilitation of cross-border investment. The forced agricultural liberalization in many developing countries, combined with the sanctioning of agricultural support policies in the North and greater protection and freedom for cross-border investment, together have weighed heavily on the shape and composition of cross-continental food chains since the mid-1990s.

One key barometer of these changes has been the consolidation of export-oriented agriculture in what have been labelled 'New Agricultural Countries', or NACs (pre-eminent members of which are Brazil, Chile, South Africa, Thailand and, most recently, China). Recent growth of agri-exports from these countries has differed from the traditional colonial model of agri-food exporting (dominated by commodities such as coffee, tea or cocoa), because it is based more centrally on the production of a range of higher-valued foods for Western consumers, notably farmed seafood, counter-seasonal fruits and vegetables, wine, and some processed foods. The development of export-oriented fresh horticulture sectors, in particular, has been identified as providing a leading edge example of these processes (Friedland 1994; Friedberg 2001). By the mid-1990s, export horticulture had become the third largest agri-food export from Sub-Saharan Africa, behind coffee and cocoa (Berry 2001: 137). This kind of growth was attached to the politics of international debt repayment and structural adjustment programmes (Friedmann and McMichael 1989; Mingione and Pugliese 1994: 56). During the 1990s, lenders and multilateral agencies implemented 'conditionality' provisions to liberalize arrangements that accorded agriculture a privileged and protected status (Stiglitz 2002). The 1980s and 1990s were periods in which policies were implemented throughout the developing world to promote agriculture's role in earning foreign exchange, as opposed to its role in providing a source of food for domestic, rural populations. Focusing upon these issues in Brazil and Argentina, Friedmann (1994: 270–1) and Sanderson (1986) identify the emergence of an export beef complex that was dependent upon feed grain imports from the developed world (especially subsidized product from the US) and detached from domestic food security concerns (see Pritchard this volume).

This expansion of export-agriculture has gone hand in hand with new food *import* complexes. The WTO's political sanctioning of Northern agricultural subsidies has encouraged steadily increased dependence by developing countries on temperate cereals and livestock. Post-NAFTA (North American Free Trade Agreement) Mexico provides a good example of these tendencies (also see Gertel this volume for an analysis of these processes in Egypt). Despite Mexico's status as a biological 'centre of origin' for maize, Mexican agriculturists have become increasingly unable to compete with subsidized American producers in a post-NAFTA environment (see Massieu and Chauvet this volume). Under the NAFTA, the Mexican government was required to eliminate orderly market arrangements for maize. Removal of price supports exposed *campesinos* to commodity markets controlled by the transnational grain traders, reducing real market maize prices to *campesinos* by 46.2 per cent during the period 1993–9 (*Public Citizen* 2001). The resultant agricultural restructuring associated with these developments has seen 1.75 million smallholder maize growers leave the land (Carlsen 2003). In developing country contexts, imported foods (such as maize) are inserted into restructured supply channels, dominated increasingly by transnational food processing firms and retail chains.

Cross-continental food chains associated with the emergence of the NACs and the food import complexes of developing countries therefore represent material conditions of contemporary global food politics. In the current context, the WTO regime has protected politically sensitive farm interests in Northern countries, while at the same time facilitating access to developing country markets and production sites. These processes have engineered changes to the global geography of food trade, and these are now addressed.

Positioning cross-continental food chains in space

The basic dynamic of recent changes to the global geography of food trade is encapsulated in Figure 1.1. During the 1990s and into the new century, the growth of world agricultural exports significantly outpaced that of world agricultural production, implying an increase in the proportion of the world's food that is traded internationally. As displayed in Table 1.1, the majority of major food groups exhibited export volumes in 2001 that were at least one-third larger than they were in 1990 and for some (oil crops, vegetable oils, vegetables, meat) export volumes in 2001 were approximately *double* in size compared to 1990.[1] Yet alongside the growth in agri-food export volumes, agri-exporters have faced a global economic environment in which the prices they have received have fallen dramatically. As Figure 1.1 shows, in 2001 the unit value (i.e. average prices) of world agricultural trade had fallen to levels not seen since the mid-1980s.

The reasons for these trends are directly connected to the global political relations of agri-food trade, as described above. On the one hand, the extensive agricultural protectionism of Northern countries contributed to production surpluses and weak international prices for broad-acre agricultural products. During 1990–2002, soybean prices fell 45.8 per cent, wheat prices fell 44.5 per cent and corn prices fell 32.5 per cent (Australian Bureau of Agricultural and Resource Economics 2003: 3). In 2002, largely because of US subsidies that saw cotton sold on world markets at 57 per cent below

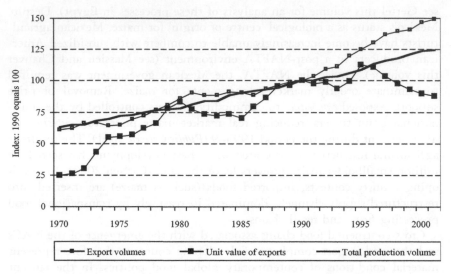

Figure 1.1 Export and production trends for agricultural products, 1970–2001
Source: WTO 2003.

Table 1.1 The size and incidence of world agri-food trade, 1990 and 2001

	1990		2001	
	Exports (000 tonnes)	Exports as a percentage of world production	Exports (000 tonnes)	Exports as a percentage of world production
Cereal grains[1]	237,200	13.35	284,112	14.90
Starchy roots	43,438	7.58	36,944	5.39
Sweeteners[2]	33,172	23.83	50,671	30.27
Pulses	7,109	12.19	10,884	20.63
Tree nuts	2,023	37.21	3,417	43.40
Oil crops	38,957	14.28	78,482	20.60
Vegetable oils	23,584	37.16	45,863	47.63
Vegetables	24,479	5.30	43,065	5.55
Fruit[3]	53,758	15.32	80,638	17.15
Stimulants[4]	10,009	87.38	13,748	96.70
Spices	711	18.17	1,285	24.93
Alcoholic beverages	12,146	6.18	19,605	8.50
Meat	14,285	7.96	25,913	10.93
Offals	786	6.39	2,065	13.34
Animal fats	6,173	19.31	6,779	21.10
Dairy products[5]	51,859	9.57	76,974	13.06
Fish and seafood	31,813	32.61	41,754	33.58

Source: FAOSTATS, World Food Balance Sheets.

Notes: (1) excluding beer; (2) mainly refined and unrefined sugar; (3) excluding wine; (4) coffee, tea and cocoa beans; (5) excluding butter, which is included in 'animal fats'.

the cost of production, real world cotton prices fell to levels not seen since the Great Depression of the 1930s (International Union of Food, Agriculture, Hotel, Restaurants, Catering Tobacco and Allied Workers Associations (IUF) 2003: 3). On the other hand, however, processes of structural adjustment and market liberalization in developing countries have encouraged considerable growth in agri-exporting as a strategy to earn valuable hard currencies. Consequently, and in conjunction with steadily rising production yields through the advent of high-input, 'green revolution' agricultures, the production volumes of many tropical commodities expanded massively during the 1990s, causing severe reductions in price levels. Between 1980 and 2000, prices collapsed for cocoa (by 71.2 per cent), coffee (64.5 per cent), palm oil (55.8 per cent), rice (60.9 per cent) and sugar (76.6 per cent) (Oxfam 2002: 151). Because of these price effects, the share of developing countries in total world agricultural exports fell from 46 per cent in 1986 to 42 per cent in 1997 (Private Sector Agricultural Trade Task Force 2002: 2). Therefore, the fundamental condition of the global agri-food system since 1990, in contrast to the period beforehand, has been a rapid expansion of agri-food export volumes, but without comparable net economic gains being accrued by agri-food exporters.

In geo-economic terms, these processes have encouraged greater divergence in the net export positions of the world's major regions. There is an increasingly stark distinction in the world's food system between major net exporters and major net importers (Figure 1.2; Figure 1.3).[2]

In general, Asia and Africa have become progressively larger net importers of food since the 1970s. Japan has become the world's single largest food importer on account of dietary transformations and the partial liberalization of domestic food policies. In the rest of East and South-East Asia, significant increases in the export of some agri-food products (such as tropical fruits) has been more than offset by the region's increasingly large appetite for imported foods connected to Western value systems (including the fast food complex and temperate fruits such as apples). The rapid growth of affluent urban middle classes in Asia has been a major driver of these trends. As seen clearly in Figure 1.2, when the East Asian economic crisis of 1997 impacted severely upon these populations, reduced import demand lead to an improvement in the region's net food export position. The transition of Africa from being a net food exporter to a net food importer reflects the conjoined effects of political, environmental and economic insecurities played out in a context of weak agricultural commodity prices. These data are testimony to the wider social crisis that has engulfed Africa over recent decades.

During this same period Latin America and Australia have become progressively larger food net exporters. The considerable expansion of Latin American net food exports in the 1990s reflects the growth of export-oriented agriculture in the region as domestic food production systems have been incorporated into the logic of international trade. According to the WTO, some 40 per cent of the increase in Latin American food exports during the

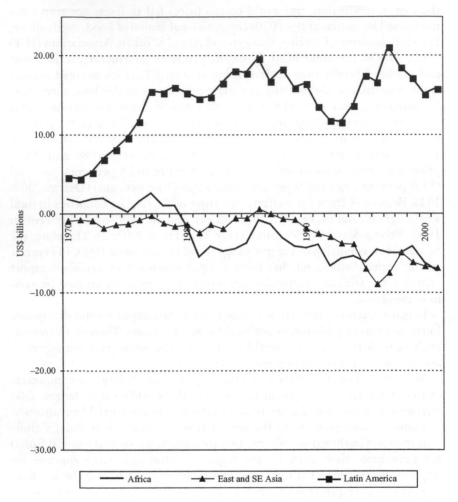

Figure 1.2 Net food trade position for Africa, East and South-East Asia, and Latin
       America, 1970–2001

Source: FAO 2003.

1990s was accounted for by exports to North America, and increased intra-
Latin American exports accounted for a further 30 per cent (WTO 2002:
185–6). Evidently, increased food exporting to North America is connected
to the advent of the NAFTA with Mexico, trade liberalization more generally and the penetration of Latin American food sectors by agri-exporting
interests attached to North American retail markets. Increased intra-Latin
American food exporting is connected to regional trade initiatives (the
MERCOSUR, ANDEAN Pact and CACM agreements)[3] and, in particular,
the rise of Chile as a key agri-exporting nation in the region. The growth

of Australian food exports is mainly the product of increased sales of raw and semi-processed products to expanding markets in the Asia-Pacific and the Middle East.

Changes to the food net export positions of North America and Western Europe also exemplify the politically constructed character of the contemporary global food system. Western Europe has progressively narrowed its

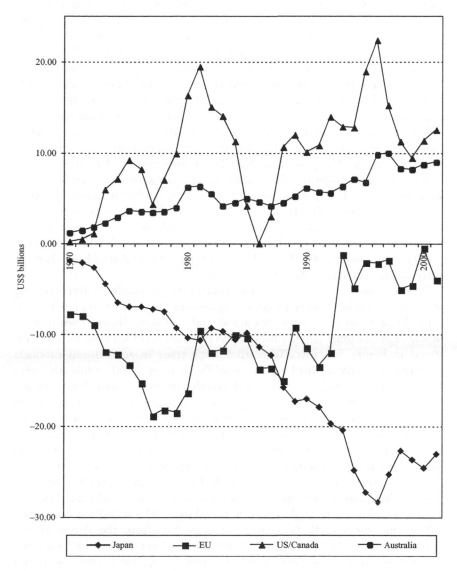

Figure 1.3 Net food trade position for Japan, the European Union, US/Canada and Australia, 1970–2001

Source: FAO 2003.

net food deficit through increased exporting: from 1990 to 2000 the value of Western European food exports increased by 35 per cent, but imports grew by only 14.3 per cent (WTO 2002: 185–6). Almost all of Western Europe's increase in food exports during this period was destined for the former Soviet bloc, Asia and North America, where advantage was taken of liberalized food import regimes and expanded food import demand.[4] At the same time, EU market access restrictions operated to limit imports from developing countries so that, for example, the value of agricultural imports from Africa increased by only 4.7 per cent during the decade (WTO 2002: 186–7). In a similar way, international trade politics fashioned the broad pattern of North American agricultural exports during the 1990s. The vast majority of the increase in North American agricultural exports since 1990 was destined for Latin America and Asia. In the case of Asia, these developments reflect the incorporation of North American agriculture within the Asian food import complex. In the case of Latin America, approximately two-thirds of the increased value of agricultural exports can be accounted for by Mexico alone. The passage of NAFTA transformed the US–Mexico agricultural trade relationship. Whereas Mexico had an agricultural trade surplus with the US in 1990, it possessed a sizeable agricultural deficit in 2002. Between 1990 and 2002 the value of Mexican agricultural exports to the US increased from US$2.56 billion to US$5.29 billion, but US agricultural exports to Mexico increased by a much faster rate, from US$2.61 billion in 1990 to US$7.06 in 2002 (United States Department of Agriculture–Foreign Agricultural Service (USDA–FAS) 2003).

When viewed in its totality, the post-1990 international restructuring of agri-food trade appears to be strengthening the role of 'regional blocs' – production and trade networks across and between adjacent continents – in the organizational geographies of the international food system. As illustrated in Figure 1.4, intra-Western Europe trade in agricultural products constituted nearly a third of the world's total in 2000, while the next regional blocs to follow – Asia and North America – trailed far behind. Moreover, the majority (about 60 per cent) of increased food trade during the 1990s was connected to increased exporting *within* continents (Figure 1.5). Intra-Western European trade, spurred on by the advent of EU economic integration, accounts for almost one-fifth of the global total increase, and intra-Asian trade (encouraged by Japan's dependence on East Asian food imports) accounts for a similar magnitude. Other significant trading relationships include integration across the Americas (two-way food trade between North America and Latin America increased by US$18.32 billion during the 1990s) and the growth of food imports to Asia from the Americas and Western Europe. In contrast, there was minimal net growth in food trade between Western Europe and North America during this period, and Africa hardly figures when a global-scale perspective is considered. What this fundamentally asserts is that the integrative logic of agri-food globalization cannot be divorced from its constituent geo-politics.

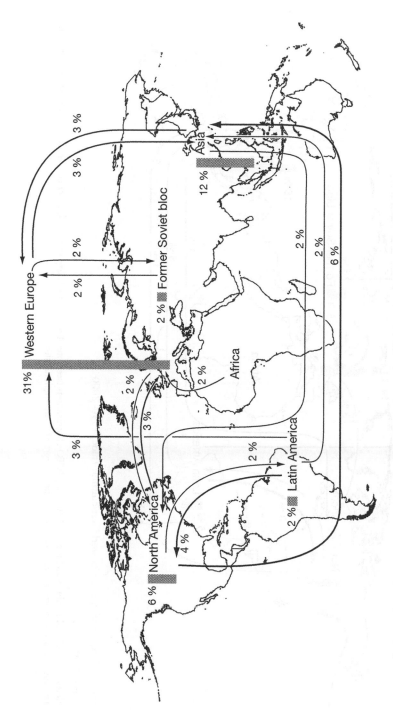

Figure 1.4 Share of world agricultural trade, 2000

Source: WTO 2001.

Notes: Agricultural products and regions follow the WTO classifications (see notes 1 and 4). Only export flows ≥ 2 per cent (rounded) of total are included. Thickness of arrows represents relative share; columns represent intra-regional flows.

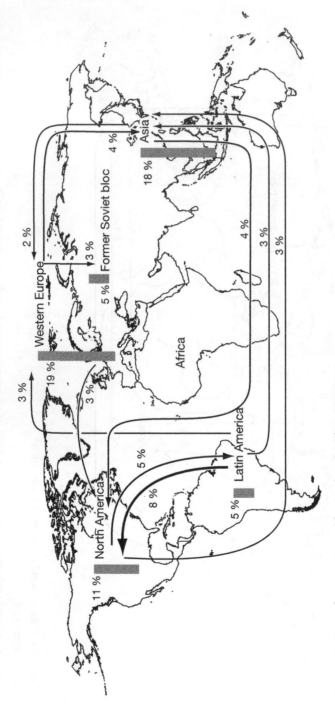

Figure 1.5 Share of increase in world agricultural trade, 1990–2000

Source: WTO 2001.

Notes: Agricultural products and regions follow the WTO classifications (see notes 1 and 4). Only export flows ≥ 2 per cent (rounded) of total are included. Thickness of arrows represents relative share; columns represent intra-regional flows.

Mapping the terminologies, concepts and directions for the analysis of cross-continental food chains

The first key agenda addressed in the main body of this book relates to the terminologies, concepts and directions for research into cross-continental food chains. Recent years have witnessed considerable debate within the social sciences on how to research the agri-food sector. There has been a fracturing of approaches and perspectives as the 'new rural sociology' method popularized in the 1980s (Buttel and Newby 1980) gave way in the 1990s to a contested research field. In his influential review article looking back over this period, Buttel (2001: 171) suggests: '1990s agrarian studies has less theoretical coherence than did early 1980s "new rural sociology"'. As Buttel explains, this was because at least six theoretical and methodological approaches vied for dominance: (i) world-historical and world-systemic analysis; (ii) global agri-food commodity chains/systems analysis; (iii) neo-regulationist studies; (iv) actor-network analyses; (v) the farming styles approach associated with the Wageningen School; and (vi) cultural-turn rural studies scholarship (2001: 171–73). Yet in drawing together the 'big picture' of how these intellectual contests have shaped this scholarly field, Buttel suggests that 'the diversification of late-twentieth-century sociology and political economy of agriculture is a good thing' (2001: 176) and that:

> Late 1990s agrarian studies is more diverse, less deterministic, more nuanced, and more anchored in empirical research than was the new rural sociology. In addition, the most recent agrarian studies literature is squarely addressing some of the key issues – the interplay of the 'global' and 'local', the society–nature dualism, homogenisation/resistance, and so on – that if anything seem destined to become more important over the next decade.
>
> (Buttel 2001: 177)

Buttel's insights hold great relevance for this book. On the one hand, its substantive thematic area – cross-continental food chains – corresponds, more or less, to one of the six theoretical/methodological approaches he flags. But at the same time, the contributions herein do not represent empirical variations of a single theoretical approach. As noted earlier in this chapter, the focus of this book is intentionally expansive. It uses 'cross-continental food chains' as a subject area from which to apply varied theoretical and method-ological approaches. As such, this book embodies the kind of diverse, nuanced and empirically grounded approach to agri-food studies suggested by Buttel in the quotation above.

This priority resonates through the four chapters in Part One of this book, which each address thematic issues. In Chapter 2, William Friedland sets forth an argument for re-conceptualizing the terminology for cross-continental food chain research. Friedland's seminal research on commodity systems analysis in the 1970s and 1980s (Friedland and Barton 1975, 1976; Friedland *et al.* 1981; Friedland 1984) was a major plank in the transformation of agrarian studies to a field of inquiry that (i) expressed its political objectives more directly and (ii) sought to explain economic processes in agriculture within the broader context of capitalist social relations. Twenty years after the publication of the most complete empirical expression of that approach – *Manufacturing Green Gold* (Friedland *et al.* 1981) – Friedland urged 'the elaboration of commodity systems methodology based on inductive methods in empirical research', through the closer examination of three key methodological areas: the scale of commodities; sectoral organization and the state; and commodity culture (Friedland 2001: 82). Friedland's chapter in this book works towards this agenda. On the basis of comparative assessment of different agri-food commodities, he asserts a need to better specify the concepts and terminologies used in agri-food research.

The ensuing three chapters in Part I then critically examine key and/or emerging themes that are integral to the future directions, shape and composition of cross-continental food chains. The interconnectivities between health regulation and the rules of agri-food trade is the subject of Chapter 3, in which David Barling and Tim Lang argue that institutional arrangements that anchor contemporary cross-continental food chains are beholden to a set of conflicts regarding the right to safe food on the one hand, and the agenda to promote liberalized trade on the other. In Chapter 4, Richard Le Heron then critiques neo-liberalism as an ideology and practice within the global agri-food sector. Observing this issue from the vantage of New Zealand – a country that, arguably, has embraced neo-liberalism above and beyond any other in the world – Le Heron identifies fundamental contradictions in the neo-liberal condition. He suggests that the practice of neo-liberalism in New Zealand is best understood as a series of intersecting political projects rather than a holistic and internally consistent meta-narrative. Chapter 5 builds on these general arguments by focusing on contradictions and political choices in the global regulation of genetically modified foods. Reviewing the recent Mexican experience, Yolanda Massieu and Michelle Chauvet expose the ways in which the spirit of the Cartagena Biosafety Protocol has been undermined by other, conflicting, WTO agreements. This case is pertinent from a global perspective, given Mexico's status as a 'centre of biological diversity'. In light of currently unresolved WTO disputes on the status of genetically modified foods, there is little doubt that the material discussed in this chapter has pressing relevance for future global directions in the organization of cross-continental food chains.

The local impacts of cross-continental food chains

The elemental political context for this book is the continuing global inequality in access to food. In 2003 some 840 million people, or about one person in eight on the planet, were chronically hungry. In the West African country of Niger, in 2001, 2,118 calories were available per person per day. This included 57.3 grams of protein and 38.8 grams of fat. By comparison, in the US during the same year 3,766 calories were available per person per day, including 114.5 grams of protein and 152.7 grams of fat (Food and Agriculture Organization (FAO) 2003). Yet hunger and food insecurity is expressed in other scales as well – for example, 31 million Americans were rated as 'food insecure' in 1999 (Andrews *et al.* 2000) – and at the same time, obesity rates are rising rapidly within both developed and developing countries.

The challenge of food inequality and hunger stalks debate on cross-continental food chains. This is seen with greatest clarity in the recent conduct of WTO negotiations, which have been premised on the need to address global inequality and development. In the wake of the abandoned 1999 Ministerial Summit in Seattle, the WTO sought to assuage its critics by professing a commitment to the agricultural and food concerns of developing countries. The multilateral trade round that commenced in 2001, in the city of Doha in the Persian Gulf nation of Qatar, included the Doha Development Agenda, a mechanism to place agriculture centrally within the negotiation process. Yet the task of putting substance to this priority has remained vexed, because of entrenched disagreement on how best to attain this goal. Some trade negotiators and advisers argue that the liberalization of agricultural markets in affluent countries, especially the European Union (EU), the US and Japan, provides the most direct route for increasing the living standards of people in the developing world. According to this line of thought, enhanced access to affluent markets will stimulate production in the developing world, and the abolition of production and export subsidies by Northern countries will inflate prices for many traded agricultural commodities. Other analysts are more sceptical over the importance of market access, suggesting that whereas it may generate increased production and exports for competitive developing country agri-food systems, these benefits will not necessarily trickle down to the population more broadly and, indeed, may be counterproductive for domestic food security because of their implications for patterns of land ownership and control (Berry 2001; Vorley 2002).

The broad debate around these issues is pursued in Part II of this book. In Chapter 6, Jane Dixon and Christina Jamieson examine recent dietary transformations in the Cook Islands, a South Pacific micro-state. Focusing on the cultural construction of poultry among Cook Islanders, Dixon and Jamieson seek to explain the (apparently contradictory) processes that have led to rapid growth in the consumption of frozen chicken meat imports in a context where free-ranging local birds are said to 'taste better' but are neglected as a food source; and in which tourists partake in supposedly 'authentic' island feasts by eating imported frozen chicken meat produced by large corporations in New Zealand, Australia or the US. This is followed, in Chapter 7, with Stewart Lockie's interpretation of Australian–Philippine food trade in the context of debates on food security. Over recent years

Australia and the Philippines have both sought to increase two-way bilateral food trade, yet as Lockie argues, this may not necessarily constructively resolve the deep-rooted problems of food insecurity within the Philippines. This chapter, therefore, provides a powerful set of arguments that bring into question dominant arguments about the relationships between trade, economic growth and food security. This focus on the relationship between trade and hunger is developed further by Jörg Gertel in Chapter 8, which explores Egypt's dependence on imported wheat. Gertel makes the point that the vulnerability of Cairo's poor to the political economy of (mainly US) wheat imports is inscribed into their bodies, via poor health. He argues for an approach to cross-continental food chain research that documents the operation of agricultural markets at an international level, and that then reaches into the 'lived conditions' of people dependent on these imported foods. Finally in this section, Sietze Vellema focuses on the social and economic arrangements that underpin the contract production of horticulture for affluent export markets. His case study of asparagus production in the Philippines for the Japanese market highlights the diverse local dynamics that can be incorporated within a single contract production scheme. Vellema's key point is that an understanding of the global political economy of contract farming needs to be built from ethnographic research practices that document the strategies of control and coordination exercised by large corporations.

Lead firms and the organization of cross-continental food chains

The historical and geo-economic conditions of the global agri-food system discussed earlier in this chapter, give rise to, and are rooted in, specific formations at the scale of individual agri-food production complexes. During the past decade extensive research has sought to document and interpret these processes, although analysts differ on the importance they ascribe to particular factors.

On the one hand, it is evident that a set of interconnected processes that are characteristic of contemporary global economic change (including the increased size, scope and purchasing power of transnational corporations and retail firms, trade liberalization, product standard harmonization and lower transport and logistics costs) appears to be encouraging a *general shift* towards greater international flexibility and production–trade coordination in agri-food complexes. In the terminology of the global commodity chains literature (Gereffi and Korzeniewicz 1994), these tendencies towards the development of geographically flexible and footloose supply chain structures are consistent with the concept of 'buyer-driven chains', where influential end-users have considerable freedoms over whom and under what circumstances they source products. Reviewing recent research on these issues, Daviron and Gibbon (2002: 152) identify a general shift towards buyerdrivenness in some commodity chains, such as coffee and cocoa, where there has been growth in the market power of large buyers and the dismantling of producer organizations, such as marketing boards. Similar developments are apparent in the tomato paste sector, which has increasingly become a standardized commodity over recent years (Pritchard and Burch 2003). Although Raikes *et al.* (2000: 399) provide the cautionary note that care needs to be taken when applying the global commodity chain concepts to the agri-food sector (the model was developed initially to account for restructuring processes in the manufacturing sector, and cannot be translated uncritically to agriculture and food), recent international agri-food restructuring has unquestionably shifted power towards those actors (notably transnational agri-food corporations and retail chains) who can exploit the advantages of geographical mobility, and, as this occurs, chains tend increasingly to take buyer-driven shape.

Transnational food corporations have been the engines for these transformations. By the year 2002, the world's ten largest food companies had a combined turnover of US$260 billion, which was the equivalent of 24 per cent of global processed food sales (Thomas 2002). The 1990s and early years of the new millennium were periods of intense merger and acquisition activity in the food sector, as leading companies jostled for market leadership. Alongside these processes, a number of large food companies based in developing countries expanded rapidly (Burch 1996, this volume; Goss *et al.* 2000; Berry 2001). Also, economic power shifted in favour of an emergent class of multinational supermarket chains, which sought to narrow supply channels and prioritize larger-scale supply systems.

Recent documentation of these developments has contributed significantly to our understanding of the dynamics involved in these restructurings of agri-food chains. Marsden's (1997) research on the São Francisco irrigated export agriculture complex of northern Brazil, the largest irrigated agricultural region of Latin America, underlines the role of near-consumer agencies, such as supermarket chains based in developed nations, in orchestrating these production complexes. Friedberg's (2001) research comparing two African-European chains for green beans (the Zambia-to-Britain chain, and the Burkina Faso-to-France chain) emphasizes that, despite vast differences in the ways these chains operate, in both cases farm producers bear the major component of risk and remain in a subservient position with respect to buyers who are generally larger and more geographically mobile. Fold's (2001, this volume) research on the West Africa–Europe cocoa–chocolate chain documents how the rise of transnational branded chocolate companies in Europe, combined with the deregulation of state marketing boards in Africa as part of structural adjustment programmes, has systemically weakened the bargaining powers of cocoa producers. These cases demonstrate that international agri-food supply complexes operate in ways which endow some actors with greater abilities to add value and exercise control than others, and that these power relations generate particular environments for risk and profit.

The chapters in Part III of this book address issues relating to these themes. In Chapter 10, Alex Hughes critically examines ethical trading initiatives undertaken by UK supermarkets. As indicated above, supermarkets have become important lead actors in the contemporary restructuring of food chains. In this chapter, she argues that the neo-liberal, self-regulatory nature of these initiatives generates tensions with supermarkets' commercial agendas. Chapter 11 then examines lead firms using a very different context. Over the past decade, the Argentinian export fruit and vegetable sector has undergone significant restructuring in the context of the entry and expansion of Expofrut/Bocchi, an Italian-owned horticultural production and trading firm. In documenting this company, Mónica Bendini and Norma Steimbreger conclude that Expofrut/Bocchi 'represents a new manifestation of the classic Latin American plantation agro-economy' because of its catalytic role in transforming production conditions within major Argentinian fruit and vegetable-growing regions. The transformative role of large companies in particular regional contexts is also a central theme of David Burch's analysis of the global poultry industry in Chapter 12. Using case studies of two international firms – the Charoen Pokphand (CP) group from Thailand and the Grampian Country Food Group (GCFG) from the UK – Burch documents how the global poultry sector has become organized via international ('North–South') corporate networks of feed, poultry and retail interests. Importantly, this chapter also documents the extensive interplay of North–South interests in this sector; CP (a 'Southern' firm) has invested in the 'South' and the 'North', whereas GCFG (a 'Northern' firm) has also invested in the 'North' and the 'South'. Burch's chapter concludes by raising questions about the role of 'nature' in the global poultry complex, given the context where 'biological' variables have been increasingly industrialized. Finally, in Part III, Charles Mather and Bridget Kenny explore the roles of lead firms as investors. Using the example of the South African dairy industry, they report the story of how two transnational dairy firms (Parmalat from Italy, and Danone from France) sought to take advantage of this sector's deregulation. The key insight from Mather and Kenny's analysis relates to the importance of historical and geographical context; the post-deregulation experiences of the South African dairy sector did not mirror those of other (so-called) emerging markets. Consequently, deregulation of South African dairy has *not* been associated with the concentration of control in the hands of multinational interests.

Multi-scalar politics and restructuring of cross-continental food chains

The world is trading more food, and relatively more of the world's food is being traded, than ever before. The questions of how this is occurring and whom it is benefiting are centrally relevant to global debates on the future of the world's economy. At times, the explicitly political character of these processes is visible in stark detail. Since the late 1990s there have been vigorous and direct protests by elements of civil society against the WTO. Yet often, contests and struggles over the construction and restructuring of the international food system are played out in arenas that are less amenable to media coverage. They occur in obscure committees that determine food standards and trade rules; behind the closed doors of financial institutions and corporate offices; in the fields and factories where the actions of workers, farmers and management shape the conditions under which food is produced, and in the hearts and minds of consumers worldwide.

The increasing complexity of these struggles over economic benefits and social justice is, not least, a result of the gradual but comprehensive dismantling of state regulation of traditional export crop production in developing countries. New forms of direct sourcing of agricultural commodities by transnational agri-food companies are being tested and consolidated, often incorporating agricultural producers into various types of contract farming. Northern quality conceptions and consumer concern for environment and labour conditions are increasingly important within the direct sourcing strategies, often secured via cooperation between private companies and NGOs taking up a function as mediators (and supervisors) between the producers and the industrial consumers.

In the final part of this book, four chapters explore different governance dimensions of these transformations. In Chapter 14, Jeffrey Neilson documents the political economy of the export of coffee from the Indonesian island of Sulawesi to Japan. The thrust of Neilson's analysis relates to the complex politics of 'Geographical Indications'. In this industry, geographies of production are integral to perceptions of quality and, therefore, price. Hence, the control of 'place' translates to the construction of value. As Neilson reports, the present private regulation of this cross-continental chain is not conducive to the retention of value by coffee growers in Sulawesi. Attention to the question of 'who writes the rules' of trade is also developed in Chapter 15, where Robert Fagan critiques a trade dispute between Australia and the Philippines regarding bananas. As Fagan argues, a close reading of this dispute brings into focus the way it has been constructed by different parties in different ways, to serve different ends. Hence Fagan suggests that a multi-scalar approach is required if the full nuances of the dispute are to be understood. In Chapter 16, Fold documents the shift towards private regulation of the West African cocoa industry, as an entry point to critique debates on the management of 'quality' in food chains. In this industry, major chocolate companies have recently collaborated in order to fund Western NGO implementation of programmes that ensure that cocoa of adequate quality is grown without the use of child labour. Fold questions why these initiatives have been developed, and what their implications are for food chain structures. Finally, in Chapter 17, Pritchard challenges globalization discourses in the beef sector. He argues that the global geography of beef production and trade remains organized by political deals and restrictions on trade orchestrated by national interests. Focusing on Australia–Japan beef trade, he traces these politics to patterns of investment and disinvestment. As the final chapter in this book, Pritchard's analysis takes us back to the fundamental argument introduced in the introduction to this chapter; namely, that 'cross-continental food chains' are not an inevitable market outcome but are politically constructed economic and social formations.

Conclusion

The construction and restructuring of cross-continental food chains are expressions of how contemporary political contests over food intersect with the historical, place-based and biophysical attributes of particular commodity complexes. The case studies of this book represent narratives on the grounded politics and economics underlying global agri-food governance. By presenting evidence on the particularities of restructuring in specific agri-food chains, this book is a vehicle for shedding light on how, in what ways and to whose benefit the global food system is being restructured. As more and more of the world's food moves across national boundaries, the task of understanding these processes becomes evermore challenging, and the debates addressed herein gain increasingly pressing relevance.

Notes

1   One of the complications in these analyses is the use of comparable data. Data in Figure 1.1 are sourced from the WTO and represent 'agricultural products'. This includes raw and processed foods but also non-food agricultural products such as fibre (cotton, wool etc.) and forest products. This is slightly different from Table 1.1, the data of which is sourced from the FAO.
2   Note that the data in these figures measure 'food and live animals' exports as estimated by the FAO. First, there are obvious difficulties with the abilities of statistical agencies in some countries to collect these data, and so these figures should be treated as indicating general trends only, as opposed to authoritative accounts. Second, there are discrepancies between this FAO series and the WTO's export database measuring 'food' (WTO 2003). Because the FAO database has a longer timeframe, it is used to construct Figures 1.2 and 1.3; however, on occasions elsewhere in the text, mention is made of WTO data. Although the FAO and WTO data series are not wholly consistent, they both highlight similar trends.
3   MERCOSUR links Argentina, Brazil, Paraguay and Uruguay into a common trade area. The ANDEAN Pact links Peru, Bolivia, Colombia, Venezuela and Ecuador. CACM is a free trade area of Honduras, Nicaragua, Guatemala, Coast Rica and El Salvador.
4   The FAO publishes data for the EU, while the WTO publishes data for 'Western Europe', which is the EU plus Turkey, Switzerland, Norway and the countries of the former Yugoslavia. Notwithstanding this difference, trends in export data for the two groups of countries are broadly similar.

References

Andrews, M., Nord, M., Bickel, G. and Carlson, S. (2000) *Household Food Insecurity in the United States*. Washington (DC): Economic Research Service, US Department of Agriculture.

Australian Bureau of Agricultural and Resource Economics (2003) *Grains Industry Performance and Outlook*. Canberra: ABARE.

Berry, A. (2001) 'When do agricultural exports help the rural poor? A political-economy approach', *Oxford Development Studies* 29 (2): 125–44.

Burch, D. (1996) 'Globalized agriculture and agri-food restructuring in Southeast Asia: the Thai experience'. In D. Burch, R.E. Rickson and G. Lawrence (eds) *Globalization and Agri-Food Restructuring: Perspectives from the Australasian Region*. Aldershot: Avebury: 323–44.

Buttel, F.H. (2001) 'Some reflections on late twentieth century agrarian political economy', *Sociologia Ruralis* 41 (2): 165–81.

Buttel, F.H. and Newby, H. (eds) (1980) *The Rural Sociology of the Advanced Societies*. London: Croom Held.

Carlsen, L. (2003) *The Mexican Farmers' Movement: Exposing the Myths of Free Trade*. Silver City (NM): Americas Program Policy Report, Interhemispheric Resource Center.

Commission of the European Communities (2002) *Mid-Term Review of the CAP*. Brussels: European Commission.

Daviron, B. and Gibbon, P. (2002) 'Global commodity chains and African export agriculture', *Journal of Agrarian Change* 2 (2): 137–61.

Fold, N. (2001) 'Restructuring of the European chocolate industry and its impact on cocoa production in West Africa', *Journal of Economic Geography* 1: 405–20.

Food and Agriculture Organization (FAO) (2003) FAOSTATS database. Online: www.fao.org (accessed various dates).

Friedberg, S. (2001) 'On the trail of the global green bean: methodological considerations in multi-site ethnography', *Global Networks*, 1 (4): 353–68.

Friedland W.H. (1984) 'Commodity systems analysis: an approach to the sociology of agriculture', *Research in Rural Sociology and Development* 1: 221–35.

Friedland, W.H. (1994) 'The new globalization: the case of fresh produce'. In A. Bonnano, L. Busch, W.H. Friedland, L. Gouveia and E. Mingione (eds) *From Columbus to ConAgra: The Globalization of Agriculture and Food*. Lawrence (KS): University of Kansas Press: 210–31.

Friedland, W.H. (2001) 'Reprise on commodity systems methodology', *International Journal of Sociology of Agriculture and Food*, 9: 82–103.

Friedland, W.H. and Barton, A. (1975) 'Destalking the wily tomato: a case study in social consequences in Californian agricultural research', *Department of Applied Behavioral Sciences, College of Agricultural and Environmental Sciences, University of California at Davis, Research Monograph Series*, 15.

Friedland, W.H. and Barton, A. (1976) 'Tomato technology', *Society*, 13: not paginated.

Friedland, W.H., Barton, A. and Thomas, R. (1981) *Manufacturing Green Gold* Cambridge: Cambridge University Press.

Friedman, H. (1994) 'Distance and durability: shaky foundations of the world food economy'. In P. McMichael (ed.) *The Global Restructuring of Agro-Food Systems*. Ithaca (NY): Cornell University Press: 258–76.

Friedman, H. and McMichael, P. (1989) 'Agriculture and the state system: the rise and decline of national agricultures, 1870 to the present', *Sociologia Ruralis* 29: 93–117.

Gereffi, G. and Korzeniewicz, M. (eds) (1994) *Commodity Chains and Global Capitalism*. Westport (CT): Greenwood Press.

Goss, J., Burch, D. and Rickson, R.E. (2000) 'Agri-food restructuring and Third World multinationals: Thailand, the CP Group and the global shrimp industry', *World Development*, 28 (3): 513–30.

International Union of Food, Agriculture, Hotel, Restaurants, Catering, Tobacco and Allied Workers Associations (IUF) (2003) *The WTO Cancun Agenda: Undermining Decent Work in Agriculture*. Sydney: IUF Asia-Pacific Secretariat.

McMichael, P. (1998) 'Global food politics', *Monthly Review*, 50 (3): 97–122.

Marsden, T.K. (1997) 'Creating space for food: the distinctiveness of recent agrarian development'. In D. Goodman and M. Watts (eds) *Globalising Food*. London: Routledge: 169–91.

Mingione, E. and Pugliese, E. (1994) 'Rural subsistence, migration, urbanization, and the new global food regime'. In A. Bonnano, L. Busch, W.H. Friedland, L. Gouveia and E. Mingione (eds) *From Columbus to ConAgra: The Globalization of Agriculture and Food*. Lawrence (KS): University of Kansas Press: 52–68.

Oxfam (2002) *Rigged Rules: Double Standards*. Oxford: Oxfam.

Pritchard, B. and Burch, D. (2003) *Agri-food Globalization in Perspective: International Restructuring in the Processing Tomato Sector*. Aldershot: Ashgate.

Private Sector Agricultural Task Force (2002) 'Supporting document to the Agricultural Trade Taskforce Communique', *World Food Summit of the FAO*, 10–13 June.

Public Citizen (2001) 'Down on the farm: NAFTA's seven-year war on farmers and ranchers in the U.S., Canada and Mexico', *Global Trade Watch*, 17: 24.

Raikes, P., Jensen, M.J. and Ponte, S. (2000) 'Global commodity chain analysis and the French *filiere* approach: comparison and critique', *Economy and Society*, 29 (3): 390–417.

Roberts, I., Podbury, T., Freeman, F. and Tulpule, V. (2001) 'A vision for multilateral agricultural policy reform', *ABARE Current Issues*, 01 (2).

Sanderson, S. (1986) 'The emergence of the world steer: internationalisation and foreign domination in Latin American cattle production'. In F.L. Tullis and W.L. Hollist (eds) *Food, The State and International Political Economy*. Lincoln (NB): University of Nebraska Press: 123–48.

Stiglitz, J. (2002) *Globalisation and its Discontents*. New York: W.W. Norton & Company.

Thomas, J. (2002) 'The giants' appetite for small fry: top 100 manufacturers, global food outlook', *Food Engineering and Ingredients*, 4 (27): 14–17.

United States Department of Agriculture–Foreign Agricultural Service (USDA-FAS) (2003) *US Trade Internet System*. Online: www.fas.usda.gov/ustrade (accessed 1 December 2003).

Vorley, B. (2002) *Sustaining Agriculture: Policy, Governance, and the Future of Family-Based Farming*. London: International Institute for Environment and Development.

World Trade Organization (WTO) (various years) *International Trade Statistics*. Geneva: WTO.

Part I

Mapping the terminologies, concepts and directions for the analysis of cross-continental food chains

2 Commodity systems

Forward to comparative analysis

William H. Friedland

Introduction

Globalization has taken on a substantial life during the past two decades. It is now being applied to such diverse topics as trade, politics, political economy, labour, corporations and communities, to name only a few.[1] With regard to the agri-food sector, the concept relates to the extension through space and time of commodity production and consumption. In horticulture, for example, it involves, among other things, the extension of growing seasons through varietal development and the establishment of new locations capable of production, notably in counter-seasonal contexts. (Thus, the southern hemisphere has become an important production location for northern hemisphere markets.) Such spatial extension of production–consumption systems over enormous distances has major integrative implications for socio-economic and cultural processes, because production–marketing links, especially for perishable commodities, require meticulous logistical integration. The fragility of many commodities and the requirements of food safety imply the re-regulation of production systems in the name of 'quality' by the large retailing chains that interface between producers and consumers (Marsden and Arce 1995: 1274).

In these contexts, this chapter is concerned with addressing three issues fundamental to debates on agri-food globalization: (1) the uneven development of commodity systems; (2) comparative commodity organization and regulation, and (3) the terminology used for conceptualizing commodity systems, chains, and *filières*. The primary methodology of the chapter is comparative – comparisons *between* and *within* commodity systems – with the ambition of making globalization accessible to empirical analysis.[2] Its basic purpose is to clarify some of the variability and misunderstanding that has accompanied recent analysis of cross-continental food chains, especially in their North–South dimensions.

The uneven development of agri-food commodity structures

Although this chapter is concerned with the agri-food sector, it is useful to begin by briefly considering a non-food commodity to establish a comparative frame of reference: automobiles, a cluster of commodity chains that is probably among the most globalized of sectors. Less than a dozen companies make millions of automobiles annually, and these companies are increasingly inter-owned by each other. Considering the enormity of the global market, a relatively small number of base models are produced on various continents for national and global markets. Variations are made for functional and marketing purposes, but many tend to be 'on the skin' rather than fundamental to product design and manufacture. Marketing strategies vary by region and nation just as they do by income, age and education.

Whereas the automobile sector (and, indeed, many other areas of manufacturing) provides certain insights that are relevant for agri-food studies, a key difference is that agriculture is intrinsically dependent on *biological* processes. This difference explains why agriculture has been delayed, in comparison with manufacturing, in responding to the dynamic processes of capitalism: economic concentration; the expansion of the proletariat; and, in recent decades, globalization. Only a few agri-food commodities possess truly globalized processes of market exchange (wheat provides an example), and even in these cases, global processes tend to relate to only some of these commodities' supply chain segments. As a general rule, the more globalized agri-food segments are near-consumer activities (supermarket retailing and distribution), some parts of food processing and certain agri-input sectors (farm machinery and agri-chemicals, for instance). Other components in these systems (notably farming) tend more commonly to be organized at regional, national or local scales (Watts and Goodman 1997: 14).[3] What these differences demonstrate is the uneven development of agriculture.

The distinctive character of agriculture – the social organization of a biological system – was early noted by Marxists. In particular, Kautsky (1988 [1899]) became preoccupied with the class composition of farmers who were not responding (like workers) to the spreading growth of capitalist industry. As industry burgeoned in Britain, Germany, France and in what later became known as the 'first world', agriculture was being left behind.[4]

Although unconcerned with agricultural developments, the comparison of agriculture with industry would have benefited from the theoretical explorations of Leon Trotsky's twinned concepts of uneven and combined development. These originated when revolutionary theorists sought to explain why the Russian working class manifested greater revolutionary potential at the beginning of the twentieth century than the earlier, larger, and more industrially conscious and organized British and German working classes. Trotsky, the proponent of the theory of uneven and combined development, wrote:

Unevenness, the most general law of the historical process, reveals itself most sharply and completely in the destiny of the backward countries. Under the whip of external necessity their backward culture is compelled to make leaps. From the universal law of unevenness thus derives another law which . . . we may call the law of *combined development* – by which we mean a drawing together of the different stages of the journey, a combining of separate steps, an amalgam of archaic with more contemporary forms.

(Trotsky 1937: 5–6, emphasis in the original)

Whole societies, social formations and *commodities* have their own unevenness and combined development. While agri-food constitutes a distinct economic sector, it is a sector composed of a large number of distinctive commodities, some of which have a close kinship to others while some are very different. One distinction, for example, is between grains and other storable commodities, as opposed to those that are 'fresh' or perishable. And within these major categories, different commodities have notably different 'life experiences'; hence, unevenness in globalizing processes will be manifested between agri-food commodities.

Thus, comparative analysis can be useful for understanding the character of agri-food commodities. To illustrate this, brief consideration is given to six agri-food commodities: processing tomatoes, fresh tomatoes, lettuce, fresh table grapes, raisins and wine.[5] In the early 1800s, most people in the Western world still produced much of their own food. Of the commodities listed above, only wine was being subjected to proto-industrialized processes of production and marketing, and then only to a fraction of the population.[6] One hundred years later, at the beginning of the twentieth century, wine had become a significant commodity in Western Europe and the US (Loubère 1978; Pinney 1989). Raisins, like other dried fruits, had already entered commodity circuits in the US, whereas fresh table grapes had not yet emerged as a significant commodity form (except in California and a few other locales). Tomatoes were still considered poisonous by much of the North American population although some localized seasonal production and distribution foreshadowed the emergence of tomatoes as a substantial commodity (Levenstein 1985). Lettuce had not yet become a national commodity and was available only locally and seasonally. Beginning in the 1920s, lettuce emerged as a national US commodity and consumption expanded more or less continuously for the remainder of the century, both in the US and Western Europe.

Therefore, one hundred years ago, as far as the majority of the US population was concerned, only two of the six agricultural products – wine and raisins – could be considered *national* commodities (albeit in limited terms when compared to the agri-industrial systems of today). The other four had not yet achieved that status. Twenty years later, lettuce was on the way to becoming a national commodity. By mid-century all six had emerged in full-blown commodity form. Three of the commodity forms – raisins, wine and processing tomatoes – could be differentiated from table grapes, fresh tomatoes and lettuce. Tomato processing technology became stabilized as the tomato 'revolution' got under way in the middle of the century so that processing tomatoes became fully commodified.

This comparative perspective illustrates the uneven development of agri-food commodities. The 'life history' of each underlines the sequential processes by which each grows, develops and expands. An understanding of these different stages of development – in Trotsky's terminology, their *combined development* – thereby helps reveal the manifestations of industrialized capitalism in the agri-food sector (Friedland *et al.* 1982: Chapter 2).

Commodity regulation and organization

The uneven development of agri-food commodities is twinned with diversity in commodity regulation and organization. Focusing on the Californian experiences of the six commodities described above highlights these differences.

In the US, food scandals during the early part of the twentieth century began a gradual process of federal food regulation through inspection (popular attention to these issues was initiated by Upton Sinclair's *The Jungle* (2002 [1906])). Nevertheless, federal and state policy left agriculture largely unregulated until the prolonged crisis that followed the US agricultural 'golden age' of 1890–1920. The crisis spurred a debate about farm prices, organization and regulation, which shifted to legislative action with the election of Franklin D. Roosevelt in 1932. In agriculture, where prices were depressed because of overproduction, the Roosevelt administration sought to regulate production and thereby create conditions under which prices would rise. Legislative action dating from Roosevelt's 'New Deal' regulated – in varying degrees – five of the six commodities examined here. This took the form of marketing orders (MOs) at federal and state levels. Marketing orders drive commodity organization. They establish rules of commodity behaviour, giving growers (and sometimes packers and shippers) the power to tax production, and (1) opening the possibility for legal controls on production flows (in the case of federal marketing orders), or (2) providing funding mechanisms for marketing and scientific research (in the case of state marketing orders) (Frank 1980; Friedland and Haight 1985).

Raisins were the sole commodity of the six that organized a federal MO controlling flow to the market (in turn, which saw the establishment of the Raisin Administrative Committee [RAC]). Prior to the formation of the RAC, raisin producers shipped product to market whenever producers desired. This led to price fluctuation when growers with little capital resources sold raisins as quickly as possible to raise cash. Since most producers were small with unsteady incomes, the flood to market at the end of the season led to price collapses. The RAC's legal control over the flow to the market curtailed this problem and tended to stabilize prices. Accompanying the federal order was a state MO, which oversaw the establishment of the California Raisin Advisory Board (CalRAB), with responsibilities for industry-based marketing. Two raisin organizations were important adjuncts of the RAC: Sun-Maid, a processing and marketing cooperative, and the Raisin Bargaining Association (RBA), also organized as a cooperative. Until the 1980s, Sun-Maid and the RBA encompassed 80–90 per cent of raisin growers. Sun-Maid set prices to its members through its grower-elected board, whereas the RBA served as bargaining agent, representing its grower members in negotiations with private raisin packing companies.

The wine industry did not organize a federal MO, but initiated a state MO, permitting the establishment of the Wine Advisory Board (WAB). The WAB was empowered to tax every gallon of wine produced in California for legislative activity and scientific research. Because states controlled alcohol distribution after Prohibition, each state produced varying regulations on wine distribution. With 48 states each setting their own rules, and with some states allocating jurisdiction to lower level government units such as counties and cities, the wine industry confronted a regulatory nightmare. WAB funds were used to hire lobbyists to standardize legislation. Funds were also used to support researchers in the Department of Viticulture and Enology at the University of California, Davis. The key wine organization was the Wine Institute, a private trade organization, which effectively controlled the WAB from 1938 until 1975.

Organization and regulatory aspects of MOs for table grapes, lettuce and processing tomatoes were much weaker than either raisins or wine, and there has never been an MO for fresh tomatoes. Table grapes had a state MO, the California Table Grape Commission (CTGC), augmented by a private trade organization, the California Grape and Tree Fruit League (CGTFL). The CTGC was primarily concerned with public relations, while the CGTFL monitored transportation costs. Both became involved whenever legislation was proposed that was considered to be inimical to the table grape industry. Both organizations also were extremely active during the late 1960s and early 1970s, when the organizing activity of the United Farm Workers (UFW) union was seen as threatening.

Lettuce and processing tomatoes each had very weak MOs restricted to agronomic research. There have been, however, powerful private trade organizations. Lettuce, for example, spawned a host of regional grower-shipper organizations to deal with specific and limited problems, often focused on labour issues. The Western Growers Association (WGA) drew in most of the larger lettuce grower-shippers and was fairly effective in countering the UFW organizing drives led by Cesar Chavez in the 1960s. Processing tomatoes had a marginal MO focused on agronomic problems in tomato production and a somewhat stronger growers' association, the California Tomato Growers Association (CTGA).

How should this variation in commodity organization and regulation be accounted for? In the case of wine, the considerable organizational and regulatory density has been an artefact of its being considered a 'sinful' product, its great capital intensity and its importance as a governmental revenue source. Making wine requires significant capital resources, and wine that makes claims for quality on the grounds, for example, of aging, increases capital requirements substantially. The peculiar regulatory status of wine, with federal and many different regulations based on each state's legislation, has raised demands for extensive industry involvement in governmental relations.

Raisin production also involves significant capital resources, albeit not as great as wine. Raisin growers must wait three years after planting before they have any grapes to dry. However, in this industry, price instability would seem to be the key determinant of regulatory and organizational density. Californian raisin production has been in crisis since the 1980s, following Greece's entry into the European Union. Prior to Greece's EU accession, California was able to ship raisins to Europe, which helped to deal with surpluses. Once that market was lost, chronic oversupply collapsed raisin prices (Hanson 1996: Chapter 3).

The lettuce industry has never lent itself to cooperative efforts; this industry attracted speculative growers who are notoriously competitive. The industry has been able to cooperate only in two major instances. The first was in dealing with Chavez and the UFW; the growers were vigorously opposed to union organizing and ultimately saw to the union's exit from the industry. In a second case, the industry defended through litigation their right to establish an industry information exchange cooperative, against federal government charges that such an entity breached antitrust legislation. Ironically, however, this was a pyrrhic victory; despite legal success, the industry was unable to make the cooperative work since growers were reluctant to exchange information that might be used by competitors. Similarly, the processing tomato MO has been weak, and has expressed little influence except to provide support for University of California, Davis, researchers working on agronomic problems. The growers' organization (the CTGA) acts primarily as a bargaining agent for its members, vis-à-vis the handful of powerful corporations in this sector.

Using processing tomatoes to recast concepts

The previous analysis illustrates the uneven development of agriculture, and the comparative dynamics that lead to different regulatory and organizational forms between commodities. The third enquiry pursued in this chapter relates to the terminology used for conceptualizing agri-food commodity systems. To this end, reference is made to Pritchard and Burch's (2003) pioneering global comparative analysis of a single commodity (processing tomatoes).

Pritchard and Burch's first major finding is that processing tomatoes do not constitute a single system; rather, the global industry is characterized by a number of discrete and separated systems, each of which consists of a number of subsystems:

> What passes for 'the global food system' consists of a set of hetero-geneous and fragmented processes, bounded in multiple ways by the separations of geography, culture, capital and knowledge . . . Global agri-food restructuring needs to be understood as an intricate set of processes operating at many scales, and on many levels, rather than a unilateral shift toward a single global marketplace.
>
> (Pritchard and Burch 2003: xi)

As illustrated in Figure 2.1, the global processing system is delineated by two major clusters of production (the US and the EU) and eight smaller ones. The representation of the global processing tomato sector in this diagram is extraordinarily useful because of its capacity to illuminate dominant patterns of production and trade at a global level. Moreover, as exposed by the detail narrative of Pritchard and Burch's study, individual production clusters each possess distinct characteristics:

> The world processing tomato industry consists of hundreds of thousands of farm and factory workers, tens of thousands of tomato farms, thou-sands of processing tomato factories, hundreds of specialist processing tomato companies, a dozen key transnational corporations, tens of thou-sands of individual products, brand names, trademarks and patents, and millions of consumers.
>
> (Pritchard and Burch 2003: 247)

Earlier, Friedland (2001: 82) pointed out the similarities of meanings in the terminological usage of 'commodity system', 'commodity chain', and '*filière*' and used them interchangeably.[7] However, Pritchard and Burch's contribution to this research field now suggests a need to recast the concep-tual terminology of commodity analyses. Each of the clusters considered by Pritchard and Burch share some similarities, but there are also significant differences between them. If we utilize Barndt's (2002) more modest study of what she labels a 'global commodity chain' (but which is limited geograph-ically to North America), the incongruities of conceptual language become obvious. This suggests the utility of standardizing language. In this vein, the following suggestions are proposed:

- *Filière* defines a particular commodity in its total global configuration.[8] The Pritchard and Burch processing tomato study would be character-ized as a *filière* analysis. Similarly, classic studies of potatoes (Salaman

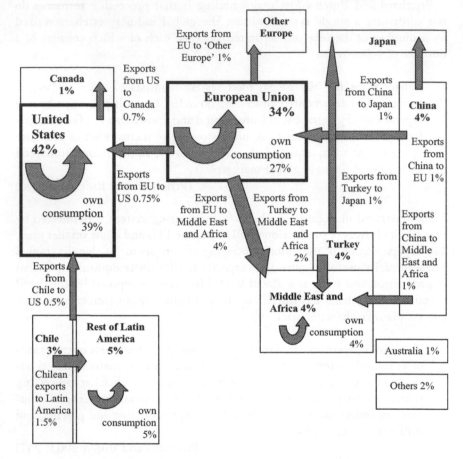

Figure 2.1 Stylized model of major production sites and trade flows in the processing
tomato industry

Source: Pritchard and Burch 2003: 253; reproduced with permission.

Note: Percentage data refers to raw product equivalents, in approximate terms only, for the years
1999–2001.

1949), sugar (Mintz 1985), and bananas (Roche 1998) would fit this
definition, notwithstanding some segments of each *filière* being ignored
in each of these works.

- *Commodity system* describes a distinct production-distribution-consump-
  tion network that is a component of a *filière*. In Pritchard and Burch's
  study, the US, the UK and Australia would each be characterized by
  this designation.
- *Commodity chain* describes a singular network of commodity production, distribution, and consumption of which Barndt's study is an
  example.

- Finally, a *segment* is a particular aspect of activity, such as growers and growing, grower organization, labour, science, distribution, marketing, culture, consumption, etc.

Thus, every production-consumption *filière* is composed of *systems* and *chains* which, in turn are composed of *segments*. Through this terminology, it becomes apparent that many widely cited agri-food commodity studies of recent decades are best described as *commodity systems* studies. These include: Dixon (2002) on Australian chickens; Friedland and Barton (1975) on Californian processing tomatoes; Friedland *et al.* (1981) on Californian iceberg lettuce; and Wright (1999) on Kentucky burley tobacco. Each of these constitutes a relatively homogeneous network but none rises to a global level. By contrast, although Barndt (2002) labels her examination of fresh tomatoes as a 'global commodity chain' study, in my terminology this is better described as a single Mexico–Canada fresh tomato *chain* study.

Each *filière*, system or chain consists of analytic segments (such as labour, grower organization, marketing, culture) that can act as the foci for analysis. Figure 2.2 sets out the twelve *segments* of the processing tomato *filière*, as presented by Pritchard and Burch. In that study, six segments are *central*: growing, first-tier processing, second-tier processing, distribution, retailing and consumption. Every chain has a number of *parallel segments* involved with all or most of the central segments: labour, capital, inputs to each segment, transportation, culture and marketing. Pritchard and Burch concentrated their analysis on the central segments, with peripheral reference to some parallel segments. The Barndt study focuses on fresh tomato central segments and includes transportation. Dixon (2002) and Wright (1999) alert us to the cultural component, which has variable importance to the central segments. In wine, for example, processors (winemakers) have a vested interest in encouraging the elaboration of wine culture; this has lower resonance with distributors but much resonance with some consumers.[9] While appreciating the enormous research in the Pritchard and Burch analysis of the global processing tomato *filière*, it is fair to guess that there will be few truly *global* commodity analyses. The complexities and detail are overwhelming. The virtue of the various studies that have been cited is that, taken in aggregate, they alert us to the various aspects of commodity life that can be studied.

To return to the central theme of uneven development, Pritchard and Burch expose the uneven development of the processing tomato *filière*, as well as suggesting the importance of combined development. This is most readily apparent in the EU. On the one hand it is internally uneven, with the processing tomato sectors of France, Portugal and northern Italy being characterized by more sophisticated forms of industrialized agriculture than those in Spain, Greece and southern Italy. Yet on the other hand, it is also apparent that the EU's processing tomato sector is being transformed through processes that seek to incorporate pre-existing European agri-*cultures* to suit the emerging requirements of industrialized agri-food sectors (in other words,

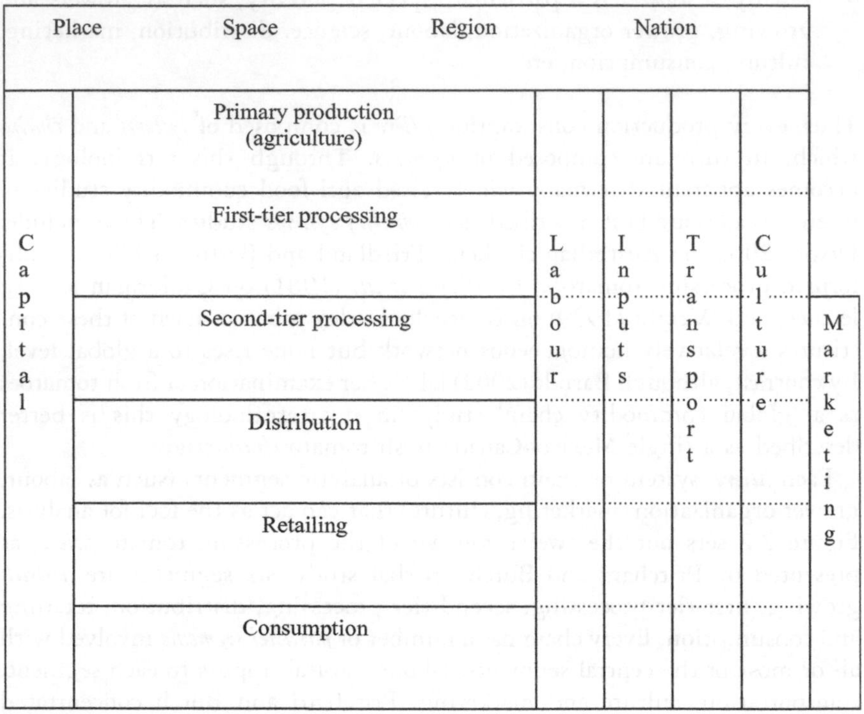

Place	Space	Region	Nation
C a p i t a l	Primary production (agriculture)	**L a b o u r I n p u t s T r a n s p o r t C u l t u r e**	**M a r k e t i n g**
	First-tier processing		
	Second-tier processing		
	Distribution		
	Retailing		
	Consumption		

Figure 2.2 Analytic map of a processing tomato segment

Source: Author.

combined development). In the EU processing tomato sector, the sizes of farms and first-tier factories tend to be smaller than equivalent segments in California. This is because Europe's processing tomato industry has a longer history than that of California, where the industry began to grow significantly only after the Second World War. The specifics of the development of the US mechanized harvesting system, driven as it was by the fear of the disappearance of cheap Mexican labour, did not have the same parallel in Europe. Moreover, Europe's Common Agricultural Policy (CAP) has been structured to give assistance to small-sized farms; in contrast to the US penchant of 'get big or get out'.

Globalization: the cross-continental dimensions

It is clear in the commodities literature that, while many studies frame their approach using globalization terminology, most rarely achieve a global level of analysis. Pritchard and Burch have provided a global analysis of what is essentially a non-global commodity. Processing tomatoes circulate minimally beyond their production region: the North American system mostly services North America; the EU system mostly services Europe and the Mediterranean, and Latin American production mainly services Latin American markets. Cross-continental competition in this sector does exist, and in certain instances is indeed intense, but this is not a primary organizational feature of the *filière*.

In other commodity analyses, there is a mix of situations: some analyses deal with commodities that are essentially global but where globality is, essentially, irrelevant. There are studies that deal with specific commodity chains between production and consumption locations, where there is cross-continental integration but no globality. And there are commodity studies that make no pretence either to a North-South dimension or globality.

Consider wine, a commodity in global circulation but where globality is only weakly relevant. Produced in a host of locations in two limited latitude belts, each nation's wine production is consumed mostly by its own population. There is South-to-North and cross-continental trade primarily from five Southern production locations – Argentina, Australia, Chile, New Zealand and South Africa – to selective Northern locations – North America and a few EU nations, particularly Britain. This trade is not insignificant but it is not of the same character as the South–North cross-continental movement of fresh fruits and vegetables where movement is critical to defining the *filière* (Friedland 1994; Freidberg 1997, 2001). Present-day counter-seasonal production and trade in some vegetables are now profound and increasing: French bean production in Burkino Faso in Africa for France (Freidberg 1997, 2001); French beans and snow peas (mangetout) from Kenya and Zimbabwe for the UK; broccoli from Central America for the US, and kiwi fruit and apples from New Zealand to the US and UK. One of the most extreme cases where South–North trade defines the *filière* is the frozen concentrated orange juice (FCOJ) sector, in which Brazilian exports play a defining role (Friedland 1991). And the historic trade in sugar and bananas, of course, has been South–North and cross-continental. But not all dominant cross-continental trade flows in particular *filières* are South–North. A considerable portion of grain movement is trans-Atlantic, augmented by some South–North movement. There is also cross-continental movement of certain horticultural products across the northern hemisphere: China has become a dominant garlic producer for the US and EU markets; The Netherlands has aggressively marketed multicoloured bell peppers (capsicum) and niche fresh tomato varieties in the US, and the US has found important markets for temperate fruits and vegetables in Japan and Hong Kong.

It is difficult to discern anything other than opportunistic patterns in most of these agri-food movements. This does not mean that there is no planning or human agency at work. In each case, local actors, usually under the goad of internal competition and encouraged by their nation-state to export, often with subsidies, seek to develop outlets for their production. Some are successful; most of those cited above are examples. Others, less well known, do not succeed. One thing is clear though: an increasing volume of agri-food products are in wider circulation globally, although very few can be said to have the kind of global spread that *filières* such as automobiles or clothing have attained.

Conclusion

If commodity studies have begun to emerge as a significant focus of analysis, it is because of the recognition of their importance in everyday life. The rush toward 'making everything everywhere' and consuming it 'everywhere' has given rise to an epidemic of studies focusing either on the macro-scale dynamics of globalization or the micro-scale dynamics of place (how globalization affects specific places at specific times). This chapter has set out an agenda for an intermediate scale of analysis focusing on commodities. Even at this level, an individual commodity analysis becomes a monumental task; hence, one of the purposes of this chapter has been to indicate ways of conducting such research.

Every commodity has a distinctive history and trajectory. These include, among other things: the availability of entrepreneurs interested in capturing wealth and status through innovations; state policies that encourage market expansion and/or scientific development; the degree to which scientific applications are made and the incentives for such scientific development; and whether a consuming population is interested in expanding food consumption inventories. The uneven and combined development of commodity trajectories is therefore an important aspect of commodity analysis. Also important is an appreciation of commodity regulation and organization. Through these foci, it is hoped that this chapter has clarified some of the variability and misunderstanding that has accompanied recent analysis of cross-continental food chains, especially in their North–South dimensions.

Notes

1   The search facility of the University of California's electronic catalogue revealed that the keyword 'globalization' turned up 14,641 hits, and the subject area 'globalization' drew 8,175 hits. This is obviously a popular topic.

2   This focus on the scale of individual commodities does *not* imply commodity fetishism. This approach is taken so that a better understanding can be obtained of the expansive character of globalisation, as well as its limitations, since not everything is being globalized.

3   Raynolds (2004: 736–9), using a commodity analysis of the globalization of organic foods, sets out 'four complementary traditions' variant from the focus of this chapter. These are useful distinctions, but because of the constraints of space they cannot be dealt with here.

4   This topic was revived in the 1980s when rural sociologists, confronted by economic concentration in agriculture, experienced as bankruptcies and the decline of family farming, queried why agriculture had resisted capitalist penetration for so long. For an early book dealing with this resuscitation, see Buttel and Newby (1980).

5 For processing tomatoes, see Friedland and Barton (1975); for fresh tomatoes see Barndt (2002); for lettuce see Friedland *et al.* (1981). Research on table grapes, raisins, and wine is ongoing and will appear in Friedland (forthcoming).

6 Although in France, this process was more advanced, with wine beginning to be consumed by a larger segment of the urban population as a commodity (Loubère 1978).

7 In addition to these terms, Fine (1994) introduced 'systems of provision', which has much the same meaning. Hendrickson and Heffernan (2002) have used 'food chain clusters' similarly, although their empirical referent is to processing raw commodities to produce food end-products.

8 While he was preceded by others (Street 1957; Goldberg 1974; Saint 1977), Lauret (1983) presented one of the earliest arguments for the study of *filières* or what came to be known in English as 'commodity systems' or 'commodity chains'.

9 Culture is variably important to agri-food systems. It is of vital importance in wine and of considerable importance, as Dixon and Wright point out, for chickens and tobacco respectively. Other commodities, in contrast, can range from tomatoes (the Heinz ketchup bottle has been called a cultural icon) to Brussels sprouts (which are essentially uncultured).

References

Barndt, D. (2002) *Tangled Routes: Women, Work, and Globalization on the Tomato Trail.* Lanham (MD): Rowman & Littlefield.

Buttel, F.H. and Newby, H. (eds) (1980) *The Rural Sociology of the Advanced Societies: Critical Perspectives.* Montclair (NJ): Allanheld, Osmun.

Dixon, J. (2002) *The Changing Chicken: Chooks, Cooks and Culinary Cultures.* Sydney: University of New South Wales Press.

Fine, B. (1994) 'Toward a political economy of food', *Review of International Political Economy* 1: 519–45.

Frank, G.L. (1980) *U.S. Agricultural Policy and the Federal and State Commodity Check-Off Programs.* Lincoln (NE): University of Nebraska.

Freidberg, S. (1997) 'Contacts, contracts, and green bean schemes: liberalisation and agro-entrepreneurship in Burkina Faso', *Journal of Modern African Studies*, 35: 101–28.

Freidberg, S. (2001) 'On the trail of the global green bean: methodological considerations in multi-site ethnography', *Global Networks*, 1: 353–68.

Friedland, W.H. (1991) 'The transnationalization of agricultural production: palimpsest of the transnational state', *International Journal of Sociology of Agriculture and Food*, 1: 48–58.

Friedland, W.H. (1994) 'The new globalization: the case of fresh produce'. In A. Bonanno, L. Busch, W.H. Friedland, L. Gouveia, and E. Mingione (eds) *From Columbus to ConAgra: Global Agriculture and Food.* Lawrence (KS): University Press of Kansas: 210–31.

Friedland, W.H. (2001) 'Reprise on commodity systems methodology', *International Journal of Sociology of Agriculture and Food*, 9: 82–103.

Friedland, W.H. (forthcoming) *Trampling Out Advantage: The Political Economy of California Wine and Grapes.*

Friedland, W.H, and Barton, A. (1975) *Destalking the Wily Tomato: A Case Study in Social Consequences in California Agricultural Research.* Davis (CA): University of California, Department of Applied Behavioral Sciences.

Friedland, W.H., Barton, A., Dancis, B., Rotkin, M. and Spiro, M. (1982) *Revolutionary Theory.* Totowa (NJ): Allanheld, Osmun.

Friedland, W.H., Barton, A.E. and Thomas, R.J. (1981) *Manufacturing Green Gold: Capital, Labor, and Technology in the Lettuce Industry*. New York: Cambridge University Press.

Friedland, W.H., and Haight, A. (1985). 'Marketing orders: a sociological perspective', unpublished paper presented at the Annual Meeting of the Rural Sociological Society.

Goldberg, R.A. (1974) *Agribusiness Management for Developing Countries – Latin America*. Cambridge (MS): Ballinger.

Hanson, V.D. (1996) *Fields Without Dreams: Defending the Agrarian Ideal*. New York: Free Press.

Hendrickson, M.K., and Heffernan, W.D. (2002) 'Opening spaces through relocalization: locating potential resistance in the weaknesses of the global food system', *Sociologia Ruralis*, 42: 347–69.

Kautsky, K. (1988 [1899]) *The Agrarian Question*. London: Zwan Publications.

Lauret, F. (1983) 'Sur les études de filières agro-alimentaire', *Economies et Sociétes, Cahiers de l'ISMEA*, 17: 721–40.

Levenstein, H.A. (1985) 'The American response to Italian food', *Food and Foodways*, 1: 1–23.

Loubère, L.A. (1978) *The Red and the White: A History of Wine in France and Italy in the Nineteenth Century*. Albany (NY): State University of New York Press.

Marsden, T.K., and Arce, A. (1995) 'Constructing quality: emerging food networks in the rural transition', *Environment and Planning A*, 27: 1261–79.

Mintz, S. W. (1985) *Sweetness and Power: The Place of Sugar in Modern History*. New York: Viking.

Pinney, T. (1989) *A History of Wine in America: From the Beginnings to Prohibition*. Berkeley (CA): University of California Press.

Pritchard, B., and Burch, D. (2003) *Agri-food Globalization in Perspective: International Restructuring in the Processing Tomato Industry*. Aldershot (UK): Ashgate.

Raynolds, L.T. (2004) 'The globalization of organic agro-food networks', *World Development*, 32 (5): 725–43.

Roche, J. (1998) *The International Banana Trade*. Cambridge: Woodhead.

Saint, W.S. Jr (1977) 'The social organization of crop production: tobacco and citrus in Bahia, Brazil', Latin American Studies Program Publications, Cornell University (Ithaca, NY): 76.

Salaman, R.N. (1949) *The History and Social Influence of the Potato*. Cambridge: Cambridge University Press.

Sinclair, U. (2002 [1906]) *The Jungle*. New York: Modern Library.

Street, J.H. (1957) *The New Revolution in the Cotton Economy: Mechanization and its Consequences*. Chapel Hill (NC): University of North Carolina Press.

Trotsky, L. (1937) *The History of the Russian Revolution*. New York: Simon & Schuster.

Watts, M., and Goodman, D. (1997) 'Agrarian questions: global appetite, local metabolism: nature, culture, and industry in fin-de-siècle agro-food systems', in D.G. Goodman and M. Watts (eds) *Globalising Food: Agrarian Questions and Global Restructuring*. London: Routledge, pp. 1–32.

Wright, D.W. (1999) *Turning Over a New Leaf: Socio-Economic and Political Transformations in the Burley Tobacco Commodity System*. Lexington (KT): University of Kentucky, Department of Sociology.

3 Trading on health

Cross-continental production and consumption tensions and the governance of international food standards*

David Barling and Tim Lang

Introduction

The setting of food standards has become a key feature in the changing contours of cross-continental food chains. In commercial trade, both private and public forms of governance have emerged in the development of food standards. Northern country governments, in particular within the European Union (EU), are setting higher levels of food standards, both for public health reasons and to underpin public confidence. But additionally, Northern corporate purchasers (notably retailers but also manufacturers and caterers) are also demanding increasingly high specifications and standards of imported fresh foods and for ingredients for food processing and manufacturing. The emphasis is on higher food standards to meet the perceived preferences of the affluent consumer markets of developed countries for quality, notably in terms of food safety. These standards are voluntary, but their observance is often mandatory for producers to gain export contracts.

Further to these developments, governments are seeking to harmonize international food standards under the trade agreements of the World Trade Organization (WTO) such as the Sanitary and Phytosanitary (SPS) agreement. The drive towards harmonization has highlighted the important role of the key international food standards setting institutions. Foremost among these is the Codex Alimentarius Commission, which was established jointly by the United Nations' (UN) Food and Agriculture Organization (FAO), and the World Health Organization (WHO). The policy debates around the setting of international food standards by these institutions reveal tensions between the goals of trade facilitation and the protection of public health.

* This chapter is based on research funded by the UK Department for International Development (DFID) in 2003 for the benefit of developing countries. The views expressed are not necessarily those of DFID. The research included 20 in-depth, semi-structured, elite interviews carried out in early 2003 with international and national food safety officials, development officials, industry representatives and international consumers' organizations. The data from these interviews have been drawn upon in the writing of this chapter.

Frequently, these issues are cast within the terms of the neo-liberal trade paradigm, emphasizing rural poverty reduction through commodity-export-led economic growth, and focusing on the effects of standards for developing country agri-food exports (Henson *et al.* 2000; Department for International Development (DFID) 2002; Jaffee 2003). Development officials and developing country governments often criticize the escalation of food standards by developed countries on the grounds that these act as non-tariff barriers to imports from developing countries. The EU, in particular, is criticized for introducing standards that are both higher and differentiated from other affluent nations. Evidently, these criticisms are grounded in an important economic truth; the EU provides an important destination for developing country produce, for instance accounting for 85 per cent of Africa's exported agricultural products by value (Commission of the European Communities (CEC) 2003a).

At the same time, however, little attention has been paid to the role of food standards in protecting the health of domestic food consumers in the developing countries. The drive towards higher food safety standards (including traceability along food supply chains) is intended to meet the needs of affluent markets and their consumers. But what of the consumers left in the developing countries? What efforts are being made to incorporate their needs in the setting of food standards? Put somewhat crudely: are the developing countries to be left with the food produce that is not deemed worthy for export?

Hence, a web of policy tensions and market signals is being generated within the governance of international food standards. A simplified diagrammatic representation of the play of these tensions is given in Figure 3.1. Fundamentally, this diagram emphasizes the interactions and conflicts between developing and developed countries; and between trade and public health. In institutional terms, the formulation of these public standards is being played out through an increasingly multilevel frame of governance. The needs of developing country consumers remain relatively marginal and lacking in advocacy within this policy web, however.

This chapter seeks to sketch out the complex interplay of policy priorities and market-led signals that emerge around the governance of international food standards setting. We examine some of the key trends occurring within the private sphere in the single European market and in the public sphere of governance, focusing on the EU and Codex Alimentarius. Finally, we identify some relatively new capacity-building initiatives that are seeking to address domestic consumption needs rather than being focused on strengthening export trade capacity.

Market-led signals, buyer-driven food supply chains and the private governance of international food standards

Multinational food manufacturers and (increasingly) retailers are at the forefront of changes to international food standards, with national government

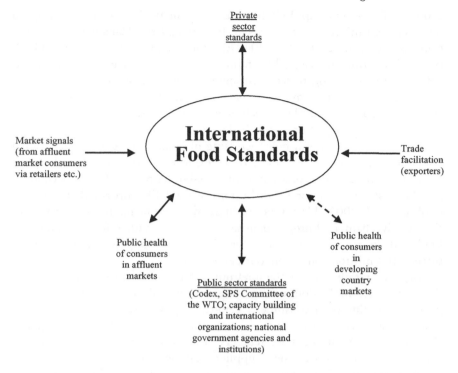

Figure 3.1 Policy tensions and international food standards
Source: Authors.

and inter-governmental bodies often lagging behind (Reardon and Farina 2001). Consequently, international food standard setting can be described as operating within a complex, bipolar regulatory structure that includes state regulatory bodies and functions at one pole and the de facto regulations stipulated by multinational food manufacturers and corporate retailers at the other. As these entities increasingly synchronize and integrate food supply chains at an international level, they exert greater influence over food standards. As such, industry-led standards increasingly are being adopted within state-based systems. This cross-hybridization of public–private standards incorporates internationally audited systems by organizations such as the European Committee for Standardization (CEN) and the International Organization for Standardization (ISO).

Food retailers have undergone rapid concentration and cross-border mergers and acquisitions in recent years (Dobson *et al.* 2003; Lang and Heasman 2004), engendering a number of changes to capacity and practices that have impacted on standards-setting processes. New retailer-led initiatives are setting up their own quasi-regulatory structures. For example, Euro-Retailer

Produce Working Group (EUREP) was set up in 1997 by 13 large European retailers to set minimum standards for Integrated Crop Management production (van der Grijp 2003: 204). EUREP's Good Agricultural Practice (GAP) protocol for fruit and vegetables followed, and has evolved from its initial defensive role in trying to set environment-friendly pesticide standards into setting standards for many more characteristics and systems (such as traceability) (EUREP 2004). There has been a rapid expansion of buying consortia and alliances among European (including UK) retailers, a phenomenon nearly two decades old but now increasing in range and scope across national boundaries (Dobson *et al.* 2003). In 1999, it was estimated that the joint turnover of the members of seven main cross-border buyer alliances accounted for about 40 per cent (or €340 billion) of total EU supermarket turnover (Dobson *et al.* 2003: 116). One business overview of retailer dominance of the supply chain in Europe identified 600 supermarket formats and 110 buying desks acting as mediators for 90 million shoppers purchasing for a further 160 million consumers (Grievink 2003).

Corporate retailers are at the forefront of buyer-driven food supply chains that are increasingly dominating cross-continental trade between African growers and European consumers. Lengthening supply chains alters power relations and who adds value and appropriates profits from these goods (Raworth 2003). In the UK, retailers are passing their quality specifications and demands on to their suppliers, giving a few specialist importers the management role of ensuring standards are met. These importers are in turn replacing traditional fresh food wholesale markets as the domestic entry point for fresh food imports (Dolan and Humphrey 2000). These supply chain trends are being witnessed across the EU to differing degrees, notwithstanding historical national regulations protecting regional wholesale markets in some member states (Gibbon 2003a). In turn, the replacement of wholesalers by fewer specialist importers is favouring contracts with larger estate producers over small-scale growers in Africa who cannot deliver the same economies of scale and the same clear traceability (Barrett *et al.* 1999; Dolan and Humphrey 2000). These standards and contract specifications are driving the standards that African growers have to meet, and are in advance of EU and national regulation. A study of Kenyan growers has found that they are rising to the challenges of meeting these higher and different standards that are being demanded, foremost by UK supermarkets but increasingly by other national retailers in the EU (Jaffee 2003).

Marks & Spencer, the UK retailer with a strong, high-value-added food presence (and seen by many as a trend setter) has unilaterally decided to phase out 79 pesticides even though some are still formally approved by the state system (Buffin 2001b). Its long-term goal is to sell residue-free produce, whereas at present 47 per cent of all produce sold in the UK currently has residues (albeit at low levels). This aspiration will require a formidable control over Marks & Spencer's 47 fresh produce suppliers who in turn work with

1,000 farmers worldwide. The Co-operative Group, with around 4 per cent UK market share, has also unilaterally banned 24 pesticides for which there are alternative growing options; six of these are still approved by the UK regulatory system (Buffin 2001a).

The retailer-driven optimization of standards is a market response to signals emanating from European consumers. The term 'user driven' has been put forward as an alternative to 'buyer driven' in order to describe the re-ordering of quality specifications by retailers and others (Gibbon 2003b). 'User driven' points to the processes where the corporate retailers seek to gain a competitive edge by meeting what they define as the signals that come from their customers. The corporate retailers research such signals exhaustively. As one leading UK retailer who was interviewed put it: 'Sometimes we have to do things before the customer even knows what they want.' That same retailer conducts a detailed tracker survey on food integrity which shows consistent rises in concerns about issues such as pesticides, genetically modified (GM) foods, additives and health issues generally. It has a pyramid or hierarchy of consumer aspirations with regard to food. At the base is the demand for food to be safe – the sine qua non of contemporary food supply management. The next most significant aspiration is healthy nutrition, followed in turn by: unnatural production; mislabelling; animal welfare; environment; and finally, ethics. Another company interviewed for this research categorizes European consumers by trends in household aspirations (e.g. seekers after hyper-convenience, authenticity, functionality, allergy avoidance, etc.).

For developing countries that seek to add value or just enter European consumer food markets, this consumer sophistication will be a key factor in mediating market entry. The mix of factors will vary over time and place. Much depends on a mix of public pressures, how consumer thinking is framed by current concerns, and how retailers and manufacturers interpret these concerns. Safety, for instance, has become the new top priority issue in the last 15 years, but prior to that the key consideration was price. Prices are still assumed today to be the key driver, but it is the other issues that distinguish between the retailers' 'offer'. The exposés of legal and illegal adulteration and poor standards have given new emphasis to safety. In the next 15 years, other concerns – such as obesity and degenerative diseases – could well replace that concern as a key driver of consumer preferences. The large retailers increasingly act as gatekeepers within international food trade. An example of their ability to act quickly to perceptions of customer concern was the rapid segregation of GM crop derivatives from their own brand produce in 1999, which was rapidly followed by many of the main branded manufacturers and mass caterers in the UK and European markets (Barling 2001). In the meantime, the move to higher food safety standards in the single European market is being reinforced by a new wave of food safety regulation from the EU's policy-makers.

The public governance of international food standards: the EU and Codex

The setting of public-sector food standards within the EU is increasingly undertaken at the intergovernmental level. National food safety legislation comes under and has to adapt to EU law. Since the late 1990s the EU has sought to strengthen and centralize food law, standards setting and risk assessment. The EU is aware of the need to meet the standards set under the WTO. The conclusion of the Uruguay Round of the GATT and the introduction of the SPS agreement and, to a lesser extent, the Technical Barriers to Trade (TBT) agreement (for labelling) signalled further intergovernmental efforts to harmonize food standards on an international basis. The setting of food standards in the public sphere is increasingly moving to intergovernmental forums. National governments and food safety authorities in Europe work within a multilevel governance frame, stretching down to regional and local authorities and upwards to EU and international regimes (Lang *et al.* 2001; Barling 2004).

The EU and food safety reform

The reform of food safety in the EU gained a high a place on its policy agenda after the Commission's mishandling of the bovine spongiform encephalopathy (BSE) crisis. In 1997 the European Commissioner Jacques Santer acknowledged shortcomings in the protection of consumer health and promised radical reform of the Commission's machinery. He called for 'nothing short of a revolution in our way of looking at food and agriculture' (Santer 1997). This set in train organizational and legislative reforms. Risk assessment processes over food and feed are now centralized under the new European Food Safety Authority (EFSA), whose mission is to 'provide scientific advice and scientific and technical support for the Community's legislation and policy in all fields which have a direct or indirect impact on food and feed safety. It shall provide independent information . . . and communicate on risks' (*Official Journal of the European Communities (OJL)* 2002: 12). The EU has undertaken a review and reform of its food law based on two key principles: a farm-to-table or whole-food-chain approach and food producers bearing primary responsibility for food (CEC 2000). The regulation on general principles of food law enacts these principles as well as defining traceability (*OJL* 2002). In addition, a regulation on traceability and labelling of genetically modified organisms (GMOs) in food and feed was passed in 2003 (*OJL* 2003).

Much of the feature of this new wave of legislation is in the form of regulation (binding on member states by means as well as ends) rather than weaker and more flexible directives (such as the legislation governing pesticides). There is a well-documented implementation deficit regarding directives that are the main body of EU environmental legislation.

Furthermore, the EU has also sought to gain greater coordination over the enforcement of food safety legislation. The legislation has focused initially on food of animal origin but in preparation are hygiene controls for foods of non-animal origin. These controls will be a prominent food safety barrier to importers in five to ten years' time and will be closely audited by the EU's Food and Veterinary Office (FVO). However, to date, inspection of imports is far from comprehensive at entry points into the single market (Jaffee 2003). Enlargement of the EU expands the entry points and coordination challenges. The new controls will include exotic fruits and fruit and vegetables, which are the key African exports to the EU. In addition, a new regulation on Official Food and Feed Controls legislation (COM (2003) 52) due to be implemented in 2005 will strengthen the role of the Commission and the FVO in inspection control. The most telling issue here is the burden of the cost for these controls, as anything over and above normal inspection will be charged to the importer. This will have important cost implications for developing countries seeking to export to the EU. Commissioner Byrne, head of the Directorate General for Health and Consumer Affairs in the European Union (DG SANCO) has stated that developing nations will have a phase-in period to adapt to new requirements on equivalence of standards, with training and twinning projects in developing countries financed with EU funds (Agra Europe 2003).

The introduction of the EFSA and the overhaul of EU food law reflected a politico-bureaucratic response to managing food safety faults in the European market and attempting to rebuild the faith of the European public. As the European Commissioners Fischler (Agriculture) and Byrne (Health and Consumer Protection) stressed in a joint statement: 'The real issue here is one of consumer confidence in the ability of the whole food chain, including public regulators, to satisfy public demand for safe quality food' (European Commission 2002). The response is clearly wide-ranging, with a further strengthening of food standards that in turn impacts upon exporters to the EU such as African growers. It adds to the diversity of standards demanded by private contractors. At the same time these legislative and institutional developments in the EU have to fall within the requirements of the WTO's international treaties on food safety and standards. These international agreements encapsulate a different mix of neo-liberal trade and market impulses, as well as trying to harmonize differing public health priorities.

The role of Codex: guardian of public health and trade?

The move towards greater international convergence of standards in the 1990s was reinforced by the introduction of a dispute resolution process on the legality of national standards under WTO agreements. In the post-WTO global regulatory environment, countries are required to base their domestic standards or technical regulations on those developed by international organizations. These organizations include: Codex Alimentarius ('Codex'), the Office International des Epizooites (OIE) for animal health and the International Plant Protection Convention (IPPC) for plant health.[1]

Prior to the completion of the GATT Uruguay Round and the SPS and TBT agreements, adoption of Codex standards at the national level was voluntary. Now there are legal obligations on countries to observe Codex standards. A member state can adopt higher levels of standards than those set by Codex in order to protect consumer health, but such actions must not be judged as discriminatory or as technical barriers to trade. Member states of the WTO can challenge a fellow member state's standards as discriminatory through the WTO dispute process. The standards, agreements and guidelines of Codex are used as a reference point in such dispute rulings. This means that the WTO has a de facto rather than *de jure* role as enforcer of standards − that is, it does not initiate disputes itself but responds to member nation complaints (Institute for Food and Agricultural Standards 2000). The threat of a dispute being invoked no doubt acts as a diplomatic lever. In addition, SPS details are discussed and communicated through the WTO's SPS committee that has become another important forum in the harmonization of food standards, albeit one with a trade focus. The observance of developing countries to SPS standards is variable although the major commodity exporters such as Brazil and South Africa are engaged with the process (Jensen 2002).

Codex's role as laid out in its articles is effectively a dual mandate of 'protecting the health of the consumers and ensuring fair practices in the food trade' (Article 1a). The balancing of public health on the one hand with trade facilitation on the other is a source of potential tension in its workings. In 2002 a Joint FAO/WHO Evaluation of the Codex Alimentarius and other FAO and WHO work on Food Standards recommended that the public health mandate be prioritized in Codex standard setting (Codex Alimentarius Commission (CAC) 2002: 7). The evaluation report's conclusions were considered at the subsequent Twenty-Fifth (Extraordinary) Session of the CAC held in February 2003 (CAC 2003). The extraordinary meeting effectively rejected this recommendation, although indicating that the issue might be revisited in the future. Participant food safety officials drew differing interpretations on the significance of this decision; some seeing it as an affirmation of the public health role of Codex, others seeing the role of trade as remaining important. Ambiguities surrounding the dual mandate remain. Specific procedural reforms to the workings of Codex were deferred at the extraordinary meeting although it was agreed that full meetings of the Codex would take place annually rather than every two years (CAC 2003).

The main workings of Codex take place in some 24 active subsidiary bodies: commodity or cross-cutting issues' committees, joint expert scientific advisory bodies or ad hoc taskforces. The process is both highly technical and slow moving. A senior European member of Codex noted the problem 'with interplay between the different committees with issues moving back and forth' between them. Decisions are highly negotiated. They go through a number of steps and are supposed to be based on consensus at each stage. To some extent, these negotiations have always been implicitly political but with the new authority given to Codex in informing standards for the high politics of international trade disputes, the politics of decision-making are becoming more explicit. Majority voting has been used more recently in some controversial areas for final approval at the full meeting of the CAC (such as the approval of growth hormones in beef and acceptance of mineral water standards). Annual meetings of the CAC are designed to speed up the decision-making but may also lead to more split voting decisions. It has been noted that the trade imperative means that Codex delegations from developing countries often contain trade not health or technical specialists (CAC 2002: 29). Developed countries dominate the committee chairs and act as hosts and secretariats to these committees, and so have some influence over proceedings. Richer countries offer bilateral aid to poorer ones to participate in committee meetings but usually have a *quid pro quo* attached. These different tensions have led to questions about the suitability of Codex to perform its role, hence the evaluation process and report in 2002. An area of particular concern has been to increase the representation and participation by developing countries in Codex and other food-standards-setting institutions. As an illustration, the 2003 extraordinary meeting of the CAC was attended by 51 out of 167 member states, of which only 20 were developing countries (CAC 2003). The issue of representation and participation is an important element that is being addressed in building the capacity of developing countries' food standards.

Food standards and capacity building for developing countries

The Extraordinary session of the CAC on 14 February 2003 voted for a new FAO/WHO Trust Fund for the Participation of Developing Countries and Countries in Transition in the Work of the Codex Alimentarius Commission (CAC 2003). The vote may be largely symbolic as initial funding promises for the new trust fund were scarce, and it joined an already crowded field of existing trust funds designed to enhance developing countries' capacity building in international food standards setting. The WTO Ministerial summit at Doha in November 2001 resulted in the so-called Doha or development round of WTO negotiations and launched the WTO/World Bank Standards and Trade Development Facility with substantial SPS elements. The FAO also has two global trust funds aimed at food safety capacity building that are seeking to raise substantial funds of US $500 million and $56 million respectively (Codex Secretariat 2002).

The EU and its member states offer a range of capacity-building supports, including interventions to remedy identified food safety system problems such as implementing new food safety systems for fisheries in African states, for example with Nile Perch exports from Lake Victoria (CEC 2003b).

The FAO provides extensive capacity building to support and strengthen national food control systems through training officials and food control staff through seminars, workshops, training manuals and guidelines and the enhancement of food laboratory facilities. In addition, the Codex Secretariat has emphasized the building up of National Codex committees and contact points involving full range of stakeholders in developing countries as a key process of SPS capacity building 'by subterfuge'. That is, it is part of a process of developing a culture change in food safety regulation, including involvement from the state, private, state enterprise and civil society sectors. Sustainability of capacity building means focusing on the domestic food safety needs of the country as well as the export market. Clearly, the private sector through the instrument of contract specification is forcing producers to raise their standards. As a UK government hosted ministerial round table pointed out from its trade-driven perspective: 'Driving up standards for local consumption would help drive up standards for export-oriented production' (DFID 2002: 13). Missing from this perspective was the public health needs of the domestic consumers in these developing countries.

The role of improving hygiene on the ground and in the local market place has been a focus of WHO capacity building, born partly out of awareness that a dual system of health standards has emerged: higher for exports, lower for local sales. This has been promoted through the Healthy Market Places initiative of the WHO African regional office. In its pioneering efforts in Tanzania and Nigeria, the initiative has sought to promote:

> the safety and wholesomeness of foods sold and traded at the markets by improving knowledge and behaviours of food vendors in food handling and sanitation. In view of the fact that poverty is largely the cause of food-borne illnesses within the region, the office integrates food safety concepts with poverty reduction activities.
>
> (Codex Secretariat 2002: 5)

The head of the WHO food safety programme has also emphasized the need to start generating data from developing countries on food safety issues to help in risk analysis and the setting of food standards. For example, it was suggested that the rapid development of risk assessment into acrylamide, a potential new class of processing-derived health risk, should be extended to developing countries to include foods prepared at high temperatures for common consumption in these countries. The Codex evaluation report also identified the need to generate data from developing countries to provide for truly global risk assessment, the current practice being to use predominantly developed country-based data for food standards setting (CAC 2002: 49).

Conclusion

The governance of international food standards is witnessing high-profile tensions between internationally harmonized standards and the raising of national standards. A complex web of policy impulses and market signals is driving the setting of international food standards. Reviewing the current private sector trends in buyer-driven cross-continental food supply chains it is clear that higher standards are market driven in Europe as retailers interpret the signals from consumers in order to gain competitive edge and so profit. Developing country exporters are meeting these standards, although the situation for producers is more demanding with evidence that suppliers are becoming more concentrated and larger among African producers. The EU is introducing both institutional and legislative overhaul of its food safety governance, following the wake of the market. The neo-liberal trade agenda has promoted international harmonization of food standards to avoid discriminatory and technical barriers to trade. However, developing countries still complain of food standards creating non-tariff barriers to their products. In the institutions of international harmonization the developing countries still feel marginal and lacking sufficient resources to participate on an equal basis. This is primarily a discourse driven by the model of poverty reduction in developing countries through export trade promotion. The protection of consumer health remains a key reason for effective food standards, as the Codex evaluation emphasized. However, the public health perspective needs to be broadened beyond the signals from affluent market consumers and citizens in the developed world in order to include developing country consumption. The safety of developing countries' consumers needs to be given more prominence in the policy making of international food standards and related capacity building, as acknowledged by the WHO and in the Codex evaluation report.

Note

1   As noted at the outset of this chapter, Codex Alimentarius was a joint initiative of the UN's Food and Agriculture Organization (FAO) and the WHO's Food Standards Programme. Its establishment in 1963 can be considered an earlier step towards the international harmonization of food standards.

References

Agra Europe (2003) *Agra Europe Weekly*, 2040, EP/2.

Barling, D. (2001) 'Supply chain perspectives – contemporary issues: the case of GM food'. In J.F. Eastham, L. Sharples and S.D. Ball (eds) *Food Supply Chain Management – A Hospitality and Retail Perspective*. Oxford: Butterworth-Heinemann: 245–56.

Barling, D. (2004) 'Food agencies as an institutional response to policy failure by the UK and the EU'. In M. Harvey, A. McMeekin and A. Warde (eds) *Qualities of Food*. Manchester: Manchester University Press: 108–28.

Barrett, H., Illbery, B., Browne, A. and Binns, T. (1999) 'Globalization and the changing networks of food supply: the importation of fresh horticultural produce from Kenya into the UK', *Transactions of the Institute of British Geographers*, 24: 159–74.

Buffin, D. (2001a) 'Retailer bans suspect pesticides', *Pesticides News*, 53, September: 3.

Buffin, D. (2001b) 'Food retailer aims to restrict pesticide use', *Pesticides News*, 54, December: 3.

Codex Alimentarius Commission (CAC) (2002) 'Conclusions and recommendations of the Joint FAO/WHO Evaluation of the Codex Alimentarius and other FAO and WHO work on Food Standards', *Alinorm 03/25/3*. Rome: Codex Alimentarius Commission, December.

Codex Alimentarius Commission (CAC) (2003) Twenty-Fifth (Extraordinary) Session, Geneva (Switzerland), 13–15 February 2003, *Alinorm 03/25/5*. Rome: Codex Alimentarius Commission, February.

Codex Secretariat (2002) *Capacity Building for Food Standards and Regulations: A report provided by the Secretariat of the Codex Alimentarius Commission based on information provided by FAO and WHO*. Rome: Codex Alimentarius Commission, 5 October.

Commission of the European Communities (CEC) (2000) *White paper on food safety, COM (1999) 719 final*. Brussels: Commission of the European Communities, 12 January.

Commission of the European Communities (CEC) (2003a) *Facts and Figures on EU Trade in Agricultural Trade: Open to Trade, Open to Developing Countries, Memo/03*. Brussels: Commission of the European Communities: 13 February.

Commission of the European Communities (CEC) (2003b) 'Lake Victoria: Commission allocates EUR30 million to support sustainable Fisheries Management', Press Releases, IP/03/86. Brussels: Commission of the European Communities: 27 February.

Department for International Development (DFID) (2002) *Ministerial Round Table on Closer Co-operation between the EU and Developing Countries on Product Standards. Report 05/02 200*. London: Department for International Development.

Dobson, P.W., Waterson, M. and Davies, S.W. (2003) 'The patterns and implications of increasing concentration in European food retailing', *Journal of Agricultural Economics*, 54 (1): 111–26.

Dolan, C. and Humphrey, J. (2000) 'Governance and trade in fresh vegetables: the impact of UK supermarkets on the African horticulture market', *The Journal of Development Studies*, 37: 145–76.

European Commission (2002) *Fischler and Byrne Final Round Table on Agriculture and Food*. EU Institutions Press Release IP/02/700, Brussels, 13 May.

Euro-Retailer Produce Working Group (EUREP) (2004) 'History – Eurepgap Fruits and Vegetables'. Online: www.eurep.org/fruit/Languages/English/index_html (accessed 5 December 2004).

Gibbon, P. (2003a) 'Value-chain governance: public regulation and entry barriers in the global fresh fruit and vegetable chain into the EU', *Development Policy Review*, 21 (5–6): 615–25.

Gibbon, P. (2003b) 'Agro-commodity chains: an introduction', paper to Overseas Development Institute (ODI), London. Online: www.odi.org.uk/speeches/gibbon.pdf (accessed 17 June 2004).

Grievenik, J.W. (2003) 'The changing face of the global food industry', presentation at the OECD conference on *Changing Dimensions of the Food Economy: Exploring the Policy Issues*, The Hague, 6 February.

Henson S., Loader R., Swinbank A., Brehdahl, M. and Lux, N. (2000) *Impact of Sanitary and Phytosanitary Measures on Developing Countries*. Reading: Centre for Food Economics Research, Department of Agricultural and Food Economics, University of Reading, April.

Institute for Food and Agricultural Standards (2000) *Markets, Rights and Equity: Food and Agricultural Standards in a Shrinking World – Recommendations from an International Workshop*, East Lansing (MI): Michigan State University, 4.

Jaffee, S. (2003) 'From challenge to opportunity: transforming Kenyan fresh vegetable trade in the context of emerging food safety and other standards', *Agriculture and Rural Development Working Paper No. 10 (Draft)*. Washington (DC): World Bank.

Jensen, M.F. (2002) 'Reviewing the SPS Agreement: a developing country perspective', *CDR Working Paper 02.3*. Copenhagen: Centre for Development Research.

Lang, T. and Heasman, M. (2004) *Food Wars: The Global Battle for Mouths, Minds and Markets*. London: Earthscan.

Lang, T., Barling, D. and Caraher, M. (2001) 'Food, social policy and the environment: towards a new model', *Social Policy and Administration*, 35 (5): 538–58.

Official Journal of the European Communities (OJL) (2002) 'Regulation (EC) No 178/2002 of the European Parliament and of the Council of 28 January 2002 laying down the general principles and requirements of food law, establishing the European Food Safety Authority and laying down procedures in matters of food safety', *Official Journal of the European Communities*, L31/1–24, 1 February.

Official Journal of the European Communities (OJL) (2003) 'Regulation (EC) No 178/2002 of the European Parliament and of the Council of 22 September 2003 concerning the traceability and labelling of genetically modified organisms and the traceability of food and feed products produced from genetically modified organisms and amending Directive 2001/18/EC', *Official Journal of the European Communities*, L268/24–28, 1 February.

Raworth, K. (2003) *Trading Away Our Rights*. Oxford: Oxfam Publications.

Reardon, T. and Farina, E. (2001) 'The rise of private food quality and safety standards: illustrations from Brazil', *International Food and Agribusiness Management Review*, 4 (4): 413–21.

Santer, J. (1997) Speech by Jacques Santer, President of the European Commission. Debate on the report by the Committee of Inquiry into BSE, European Parliament, Strasbourg, 18 February. *Speech 97/39*. Brussels: European Commission.

van der Grijp, N. (2003) 'European food industry initiatives reducing pesticide use'. In F. den Hond, P. Groenewegen and N.M. van Straalen (eds) *Pesticides: Problems, Improvements, Alternatives*. Oxford: Blackwell Science, pp. 199–220.

4 Reconstituting New Zealand's agri-food chains for international competition*

Richard Le Heron

Introduction

At the end of the twentieth century New Zealand was regarded as an important case study for impacts of an explicitly neo-liberal agenda on export-oriented agriculture. Indeed New Zealand has often been held up as a model for key elements of this agenda, especially marketization, privatization and trade liberalization. The application of the model in New Zealand rests on three key assumptions. First, liberalized trade provides the optimal environment for New Zealand's export agriculture to exploit its relatively low production costs. Second, liberalization will encourage labour and capital to shift to areas in which New Zealand has comparative advantage. Third, international competition encourages innovation, in products and institutions, helping to forge a wealthier future.

Considerable debate surrounds the legitimacy or otherwise of this political strategy, and its wider global lessons for the economic and social regulation of cross-continental food chains. In this context, this chapter scrutinizes the question of what kind of 'on-the-ground' neo-liberal agriculture has actually emerged in New Zealand. Understanding of the present political economy of the country's agri-food sector gives a better basis for assessing what responses might be expected from the sector in the foreseeable future. In this regard, the key insight developed here is that recent reworking of agri-food relations in New Zealand reflects an evolutionary state that is more than a neo-liberal ideal type. Informed by recent readings of both regulation theory (MacLeod and Goodwin 1999; Jessop 1999, 2000; MacLeod 2001) and neo-Foucaultian analysis (Barry *et al.* 1996; Dean 1999; Rose 1999), the chapter argues that the neo-liberal experiment in New Zealand agriculture can be understood most usefully by focusing first on emerging structural arrangements, and second on considering how different political projects are picked up or rejected by sector actors.

* The research for the chapter was funded in part by the Marsden Fund contract, University of Auckland, 3368388. I would also thank Wendy Larner for perceptive and critical comments on a draft of the chapter and Hugh Campbell for ongoing dialogue about transformations in New Zealand's agri-food sector.

Discussion on these issues extends the treatment of commodity chains found in the political economy of agriculture literature (Bonanno *et al.* 1994; Friedland 2001) to include both the realignment of supply chains and the appearance of 'whole-system approaches' to governance issues at a national scale. This concern for whole systems and governance underlines the activation and development of political projects in new conditions. The focus on political projects as a discursive and material dimension of commodity chain analysis strengthens the political dimensions of regulation theory as it has been applied to agri-food analysis. The emphasis on the politics of decision-making reveals the often contradictory political pressures within which agri-food actors engage. The agri-food sector in New Zealand is therefore viewed in terms of the realignment of supply chains that form the context in which different political projects are played out.

The remainder of the chapter deals with the debates on the privatized regulation of New Zealand's food and fibre chains, before specifically introducing two political projects that have sought to realign supply chains and adopt institutional techniques to encourage whole-system coordination. These two projects are New Zealand's responses to genetically modified organisms (GMOs) and its implementation of sustainable development principles in policy settings. The discussion around these cases outlines the sometimes diverging perspectives about supply chain reorganization in changing conditions and the importance of standards and benchmarks as devices for ordering system-wide and individual producer performance. The manner in which these projects have intruded upon and been encountered by agriculture suggests accounts about neo-liberal agriculture need to go beyond regulation theory to include consideration of the mobilization and use of politically inspired strategic narratives. The final section outlines the cross-continental implications of the New Zealand case. Because the two political projects selected for the chapter have strong international resonances, the New Zealand scene is, with its emphasis on supply chains and whole system coordination, illustrative of a deepening fabric of political relations in the spheres of agriculture and food.

Supply chain realignment by private interests

With state-led restructuring of agriculture in New Zealand, new conditions of production were ushered in (Sandrey and Reynolds 1991; Johnson 1992) and new space created for governing arrangements that differed from the immediate past (Le Heron and Roche 1999). This spelt the sudden end to a regulatory regime based on selling bulk agricultural commodities, with state governance aiming to maximize production volume through government-to-government negotiation. This regime had at its centre producer-marketing boards around which commodity chains were organized. New Zealand's reform strategy coincided with structural developments overseas and was further propelled by a deregulatory dynamic that impacted directly on the farm and processing sectors (Le Heron 1989a, 1989b). In combination, these processes constrained strategic choices for existing and emerging actors in what were still, in the mid-1980s, 'commodity' chains, resting on undifferentiated and traded products such as lamb carcases, butter, cheese, milk powder, apples and kiwi fruit (Le Heron and Roche 1996a). The issue for the state in the new context switched from directing coordination to facilitating emerging lines of coordination by private investors. In a little over a decade, the commodity chains had become supply chains.

Internationally, these initiatives occurred hand in hand with the rapid concentration of retail capital in the key export markets of the EU (especially the UK) and, more recently, Asia and the US. These processes have shifted the crucial cross-border regulatory site from governments to corporations, especially supermarkets (Blythman 2003; Chapter 10, this volume). In New Zealand, concentration and ownership changes also occurred rapidly. In meat-packing, the exit of foreign firms during the 1990s resulted in New Zealand interests taking a majority stake in this industry for the first time in over 150 years (Lynch 2001). Dairy mergers saw the creation of Fonterra, a near-monopoly dairy processor in New Zealand that currently accounts for 40 per cent of world dairy trade. In horticulture, Enza* replaced the Apple and Pear Marketing Board and Zespri** became the main company for kiwi fruit exports. By 2002, corporate entities had replaced producer boards for all of New Zealand's major agri-food chains (Hayward and Le Heron 2002; McKenna and Murray 2002). Analysis for the period 2000–2 indicates the added value embedded in New Zealand's agri-exports as 51 per cent of meat exports and 35 per cent each for dairy and fruit and vegetables (Bull 2003: C4). Elsewhere Lynch (2002) reports that only 5 per cent of sheep meat is exported as carcases, compared to 80 per cent in the early 1990s. Moreover, the horticultural sector continues to take advantage of 'permanent global summer time' in northern hemisphere supermarkets.

These changes in the geo-political and geo-economic agricultural trade landscape meant producers at all levels directly met international competition. In the production worlds of each actor (farmers, companies, Crown Research Institutes,[1] producer associations and so on) strategy is about many things: adding value, maximizing profits, assuring quality, defining standards, benchmarking, growing brands, meeting overseas purchasers and so forth (Le Heron and Roche 1996b; Le Heron 2003; Larner and Le Heron 2004). The basis of international competition is increasingly seen as the supply chain, with the term supply chain denoting the qualitative change from commodity to differentiated production as perceived and understood

---

*   Enza Ltd is a New Zealand food company formed in 1994 by the New Zealand Apple and Pear Marketing Board. Its primary activities involve pip-fruit exporting. In 2003 it was merged with the private-owned New Zealand horticulture company, Turner and Growers Ltd.

** Zespri Ltd is a New Zealand kiwi-fruit exporting company owned by 2,500 growers.

by an emergent and realigning cast of actors. But by whom, for what ends? And what forms of whole-system governing have begun to appear and how are they understood?

First, there was early realization that the challenges in the new environment were both comprehensive in nature and involved a different system conception. Producer boards were the object of a privatizing rationality informing neo-liberal reforms and were steadily restructured into corporate-style entities. This momentum fed back into reassessments by processors, of what might be needed in terms of changes on the farm, and by farmers, of how other chain actors might perform in farmer interests. This said, farmers were not especially well positioned to compete in the new environment. Even by the mid-1990s frustrations abounded over changing the culture of farming, which broke away from the more prescriptive culture under state intervention. The following quotes highlight the culture shift by actors that was demanded as supply chain transitions were confronted. As the Meat Producers Board (1995) *Annual Report* stated: 'It is very easy to give reasons for not doing a farm business plan. They run something like this. I'm not working if I'm planning. I know what I'm doing, it's in my head. I'm too busy.' Three years later, Chairman John Acland offered perceptive insight:

> While much has been talked about the knowledge economy, few know what it means. It is not, as we have been led to believe, a new energy economy, but *rather it is a whole new way of thinking about how we do business in our existing industries.*
>
> (Meat Producers Board 1999: 8; italics added)

Interviews with sheep meat processors during this period also confirm the magnitude of the transition, and the direction of change. One manager expressed the challenge thus: 'How do we get them (farmers) to find out how much they really understand about the game other than "I got rid of the lambs"?' Yet by 2003, outgoing Meat Industry Association Chief Executive Officer Brian Lynch (2003) was able to say in the case of the sheep meat industry:

> It is getting used to being a New Zealand industry . . . The (now) New Zealand companies, previously oriented to procurement, skipped adolescence, moving instead to adulthood, where they are seeking to differentiate themselves, through new supply chain relationships.

Second, by the beginning of the twenty-first century analysts and industry commentators found they needed to address, as a matter of course, new technical themes associated with whole-of-supply chain management. For sheep meat, supply chain thinking meant taking seriously traceability of the product back to the farm gate: 'One of the greatest challenges is linking each meat cut at the boning stage of processing, with the animal it originated from' (Meat Industry New Zealand 2002: 5). In horticulture, Bourne (2003: 30) wrote:

> I had one grower say to me, 'Europe is huge and they don't need us as much as we need them', we simply aren't in a position to pick and choose which markets we want to sell to and that means meeting the demands of our international customers.

The success of Kiwi Green (as distinct from organic kiwi fruit) and Kiwi Gold under integrated production management rests on the discipline of overseas protocols and audit, applied throughout the supply chain. Frequently, the rationale for these new agendas was connected to the need to fill the gap left by the withdrawal of government extension and support services.

A recurrent message is that the means of controlling coordination has altered, from state incentives and assistance to guide investor patterns to a centring on whole-of-system business–customer relationships. Two further quotes illustrate the changing direction and character of control. The growth of contracts in sheep meat supply has proceeded through re-establishing links between farmers and processors. The remark of a livestock representative identifies crucial aspects of realignment:

> There's been two significant targets I've had as livestock manager (1) building a good core of suitable suppliers and the other is getting hands on management of our business – getting control, because the chain is only as good as its weakest link.
>
> (Processor Manager, April 1999)

Similar sentiments guided the first joint trans-Tasman horticultural growers conference, which had a conference theme of 'Take control: strengthen the chain' (Anon. 2003a: 2).

Two further dimensions are pertinent: the importance of new performance levels, and the scale of transformations being induced. Performance at all levels emphasizes adding value. Whether at the processor or farm level, the message is clear: governing by technical means has become commonplace.[2] In the dairy sector:

> Dexcel was created to ensure the dairy industry (and in particular farmers) 'owned' the competencies necessary to achieve the four per cent annual productivity target [imposed by Fonterra]. The core competency was identified as 'the ability to manage and integrate a network of capabilities to optimise whole farm systems'.
>
> (Caradus 2003: 64)

With the new patterns of coordination, farmers, processors and purchasers are being brought closer together. Zespri believes its recent successes are built in part on 'strong results from integration of production and integration in the Zespri system – strength of the "in-market teams" direct relationships with retailers', while sheep meat farmers mention, following visits by Tesco and Waitrose to New Zealand, that it is 'nice to feel an important part of the supply chain' (farmer interview October 1998). The script in the above quotes is about knowledgeably coordinating one's activities into association with the activities of others in the chain.

This section has shown that while there is no single whole-of-system view, the disposition to explore different whole-system configurations has emerged as the New Zealand agri-food chain actors encompass increasingly different local production and overseas market conditions. The argument has stressed 'coordination' by 'private interests'. The open question is, however, 'But coordination of what kind of agriculture?' New Zealand's international prominence in the GMO and sustainable development debates has placed New Zealand agriculture once more under the spotlight. These two political projects illustrate how agriculture is (yet again) a site of moral and economic contestation. Their key relevance for this chapter is that in spite of the emergence of supply chain coordination, it is political projects that are shaping the boundaries and content of agricultural and food production. Indeed, the New Zealand scene suggests active re-engagement by the state in the governance of the agri-food sector. While a new institutional framework has been embedded, the actual nature and workings of contemporary neo-liberalized agriculture, at least in New Zealand, is very problematic.

Remapping agriculture through 'world class regulation'

The advent of genetic modification (GM) posed a particular problem for New Zealand. The GM option was widely seen to present a range of unprecedented threats to market access as well as novel production possibilities. It brought into the open the question of leadership in the world food economy. The brief discussion of this issue in this section suggests that in the face of an assumed trade-liberal world the capacity of the New Zealand government to institute a regulatory framework that contained the risks would give New Zealand a trade advantage over competitors.

New Zealand's GM story has at least three formative moments. First, a particular politics succeeded in placing GM on the national (and international) stage. As early as 1988, New Zealand's Environmental Risk Management Authority allowed GM field trials. Despite efforts a decade later by the National Government to authorize more extensive uptake in New Zealand via an Independent Biotechnology Advisory Council, New Zealand's Green Party leveraged a Royal Commission on Genetic Modification after the fifth Labour Government gained office.[3] The second moment took the form of the Commission itself, a complex and ambitious venture, involving elaborate consultative and deliberative processes. The Commission recommended the insertion of the status of 'conditional release' into the regulatory framework of the Hazardous and Noxious Substances Act. This new category enabled the Commission to portray GM as both a thinkable and a manageable development in the New Zealand context. Fundamental to this framing of GM as something that can be regulated is the remapping of New Zealand agriculture into a number of discrete segments, including conventional land-based production, integrated pest management, organic and production that utilizes GMOs (see Le Heron (2003) for a discussion of the governmental features of the Commission process). Clearly supporting GM and biotechnology as one platform for New Zealand economy and society but also attempting to buy time, the government granted an 18-month moratorium, which expired in October 2003, amid stormy protest.

Reportage in the last days of the moratorium focused on overseas experience which:

> illustrates the need for three essential elements for achieving effective co-existence of GM with non-GM production systems; a robust regulatory approach, case by case introductions and a 'whole of production chain' approach to address any identified concerns from seed production, follow-up paddock management to post-harvest handling, management and distribution.
>
> (Anon. 2003b: 9)

Some overseas observers ask:

> who holds the responsibility for monitoring separate channels for GMO and non-GMO crops; is the grain handling industry able to deal with a two-tiered delivery channel, one for GMO and the other for non-GMO; who will test the crop and at what cost?
>
> (Anon. 2003c: 43)

Labour Party literature in electorates outlined the re-regulatory intent:

> The government is following the recommendations of the Royal Commission, and is putting in place a world class regulatory system that will maximise the benefits and minimise the risks. This system will enable . . . case-by-case (assessments). Once changes are in place to the regulatory system, New Zealanders will be able to enjoy the opportunities of organic and conventional agriculture, while not closing the door to the contribution that GM may make to our way of life.
>
> (Hartley 2003: 1)

Noticeably, the GM story has not been narrated with respect to its main proponents or the benefits that the proponents and others might achieve through the approval of GM production. Biotechnology strategy covers the whole economy, being one of the foundations of the present government's growth and innovation strategy. Prospects of adding value through intellectual property in biotechnology neatly fit the mould of conventional economics, science and industry. Key promoters are the Crown Research Institutes. Upbeat comment by one of these (HortResearch) exemplifies this logic:

> The challenge for HortResearch is to reposition itself into being a globally competitive science provider ... The switch away from organophosphate insecticides has been dramatic, but the very real risks of development of resistance or product withdrawal associated with insect growth regulators and other alternative pesticides mean that the journey towards sustainability is far from complete.
>
> (HortResearch 2000: 1, 11)

If supply chains are the context in which GM is made manifest, then the regulatory surge around GM may seem paradoxical, since such institutional developments are antithetical to the espoused marketization and privatization principles of neo-liberalism. However, re-regulation should be understood as a contribution to, rather than a detraction from, such principles. First, once the general framework has been established supply chain actors will do the policing. Second, these actors will have delegated powers to be able to deploy a range of governmental techniques (i.e. standards, benchmarking, audit) and through these techniques to be able to say whether any given supply chain is GM or non-GM.

Sustainability as a metaphor of new practices in science and agriculture

This section outlines the short trajectory of a political project that has much less visibility than GM, yet has as much potential to rewrite the political and social economy of New Zealand food and agriculture. The project merits attention because in the laboratories and field stations of New Zealand's science community, the impact of a new approach to science and different expectations about what and how science might contribute to society has cut deeply into funding arrangements. The implications for agriculture are potentially profound (see Tanaka *et al.* (1999) and Kloppenburg (1988) for discussion of links between science and agri-food transformation).

Between November 2002 and August 2003, the primary science research funding agency in New Zealand, the Foundation for Research, Science and Technology (FoRST), initiated a Sustainable Development Investment Process with responsibility for allocating NZ$56 million (approximately US$30 million) per annum over the next six years. The investment process emerged from a Sustainability Review, involving the deliberations of six working groups and external commentaries (Benfell 2003: A13). FoRST then advertised the Sustainable Development portfolio and toured the country explaining its features, expectations about bids, and procedures for applications. During this process FoRST staff drew attention to two important shifts in thinking: that the science underpinning applications needed to reflect sustainability concepts; and that research outcomes were to be closely scrutinized. In its preamble to the request for proposals, the Foundation states its expectations for:

> forward-looking and opportunity-seeking investment. We are seeking research that goes well beyond clean-up and a reactive search for ways to deal with yesterday's environmental problems. We expect research teams to make full use of international thinking on the subject of sustainability and develop innovative approaches that are meaningful, effective and of priority for our own communities and businesses, in ways that are relevant to New Zealand's special social and environmental conditions.
>
> (FoRST 2002: 13)

A set of sustainability concepts was articulated: the importance of the biophysical environment; integration of environmental, economic and social dimensions; a holistic approach; a long-term view; understanding of people and values and the institutions of which they are a part; partnering with communities within the research process; and collaboration (FoRST 2002: 14). Importantly, FoRST prioritized research 'that is conducive to sustainable development as a process or pathway and transitions to increased sustainable development'.

Judging by the controversy that ensued when funding was announced (Dann 2003b; Editorial 2003; Freeth 2003), the profile of funded projects did not reproduce existing scientific structures and expectations. In particular, two areas stood out: (1) research aimed at giving greater understanding of the human and social aspects of sustainability and how all components of sustainability interact, and (2) research that was already developing partnerships and collaborations.

Several projects (amounting to over 5 per cent of the portfolio allocation) form a broad programme of research into sustainability practices (Fairweather and Campbell 2003; Manhire *et al.* 2002). They conceptualize sustainability within a value chain (read supply chain) framework, exploring issues from farming to processing, in ways never previously undertaken in New Zealand (or perhaps the world). For example, one project examines changes over time in the economic, environmental and social variables of two cohorts of farms in the four key sectors of New Zealand agriculture: farms about to convert to alternative production systems; and farms retaining status quo arrangements. Another project examines the ecological footprint in relation to on-farm and post-farm gate value chains in meat, wool, dairy and forestry, against the background of New Zealand's rating as the fourth largest consumer of biosphere resources in the world by the World Wildlife Fund. The study will elaborate features within the organizational and cultural structures that promote or prohibit decisions and behaviour that support sustainability. A third project situates sustainability in the context of tensions between democracy and sustainability, and integrating fuller costs of natural resource use into the economy. The projects are designed as collaboration among supply chain stakeholders.

The significance of these projects lies in what they displaced (three Crown Research Institutes have publicly protested about funding cuts), what they represent (a more holistic version of science bringing in social, cultural and economic aspects under the heading of human dimensions), and how they revisit value chain research in more critical and comprehensive terms (thereby breaking the compartmentalization of activities in agriculture and science). Much rides on FoRST effectively monitoring project progress and outcomes. The projects could stumble, they may embed different research practices or they might expose the tensions surrounding trade-offs in attempts to develop transitional pathways towards holistically inspired sustainability. Such projects, moreover, are not mounted in isolation. Under the banner of 'What's all the fuss about sustainability', New Zealand's Ministry of Agriculture and Food explains: 'In the last decade, key markets and their associated supply chains have required farmers to provide ever-expanding assurances about the quality of their production. Increasingly, these assurances include sustainability' (Anon. 2003d: 9). Further, the manager of Meat and Wool Innovation's quality and risk team says:

> It is important to set the scene and get everyone familiar with the concepts involved (of sustainability). Some common or shared approaches among New Zealand primary sector on how best to respond to sustainability pressures will greatly ease farmer concerns and reduce compliance fatigue.
>
> (Anon. 2003d: 11)

Yet another viewpoint suggests that: 'The policy signals to Crown Research Institutes and the universities are skewed too far towards commercialisation at the expense of applied research' (Freeth 2003: E10). Less charitably, a prominent New Zealand newspaper editorialized that the 'Foundation's allocations bear the stamp of the Government's troublesome ally, the Green Party' (Editorial 2003: A14).

The window on efforts to evaluate different models of supply chain relations is also a journey in institutional experimentation. Like the proposed GM regulatory framework, which centres on maintaining distinctly different supply chains, the sustainable development studies are to consider attributes of performance of pre-classified models of production. Thus, while the object of governance in the case of GM is the whole of New Zealand (a signal to the world that all is well), the objects of governance relating to sustainable development have yet to be constituted through the collaborative processes of new-generation science.

Conclusion

The chapter has painted a picture of major re-orderings in New Zealand's agri-food chains over the past decade. More to the point, the realigning supply chains are distinguished by several structural features found elsewhere in the world. A handful of processors and retailers dominate each chain. A consolidating set of larger farming enterprises now work the land, often in diversified, multi-farm groupings. The dynamics and patterns of realignment can be read as consistent with and confirming the neo-liberal agenda of marketization, privatization and trade liberalization. The various impulses of supply chain re-development are implicated in new patterns and re-configurations as different interests push political agenda. Uncertainties introduced into the world trading scene by the breakdown of the WTO agenda at Cancun in September 2003, or the rise of bilateralism (at the time of writing, New Zealand is pursuing a free trade agreement with China) will be assessed by supply chain actors who are presently encountering competing narratives about New Zealand's agri-food future. This is the special interest of the New Zealand case.

The political narratives of GM and sustainability represent urges to govern that run against neo-liberal tenets. Almost ironically, the neo-liberalized environment, rather than removing or restricting the arrival of political initiatives, may even be opening space for these. Importantly, both GM and sustainability are amenable to and constituted by whole-of-system coordination. The GM project as it has unfolded in New Zealand is heavily influenced by a view that the future is knowable. This reasoning is predicated on the assumption that the right questions have already been put (i.e. what is needed is a regulatory framework), so the policy task is how to find the answers (i.e. which arrangements will best contain risks to acceptable levels). In contrast, the Sustainable Development project is tentative, embracing the notion that neither the questions nor the answers are known. Khachatourians (2001: 21), a philosopher of science, argues this is science at its best, challenging preconceptions and shifting the emphasis of science away from problem solving to 'defining the unknowns and seeking new knowns, thereby creating the foundation for sound governance'.

In New Zealand the contingent collision of GM and sustainability, intersecting through links with actors in realigned supply chains, does indeed create new openings to discursively frame and perform the world differently. GM and sustainability both work with supply chains at the same time as they broaden the supply chain category to include other aspirations and actors. This has implications at the level of cross-continental agri-food chains, whether in an environment where freer trade is on the retreat or on the ascendancy. Over a decade ago New Zealand agriculture was reeling from neo-liberal restructuring. Today, the sector's complexion reflects the rapid institution of supply chain frameworks and the imprint of political debates about what sort of agriculture should be prioritized and how it should be pursued. The new cross-continental relations being forged by New Zealand's agri-food investors and producers through practices of supply chain competition increasingly embody claims and assurances about the sustainability of the sector. Thus, studies of the continuing engagement of New Zealand in the globalizing agri-food economy need to acknowledge developments after and other than neo-liberal reforms.

Notes

1   Crown Research Institutes are commercial research entities formed from the restructuring of government departments and agencies during New Zealand's neo-liberal era.
2   For example, in processing, Zespri contends, 'We are a marketing company, not a trading company . . . We aren't a commodity. We are a branded product' (Dann 2003a: C1). For farming: 'Seven thousand dairy farmers – over half the total number in New Zealand – are now routinely using the Internet to upload and download information for benchmarking and calculation on their own farms. It would be one of the highest levels of uptake among farmers in the world' (Tyson 2003: 34).
3   A 'Royal Commission' is a commission of inquiry with wide-ranging judicial powers.

Bibliography

Anon. (2003a) 'Berry growers to take control at land mark conference', *Grower*, 58 (6): July: 41.
Anon. (2003b) 'News. We go with GM', *Grower*, 58 (4): May: 9.
Anon. (2003c) 'GMO feature', *Grower*, 58 (7): August: 39–45.
Anon. (2003d) 'What's all the fuss about sustainability?', *Wool Innovation*, 3: 9–12.
Barry, A., Osborne, T. and Rose, N. (eds) (1996) *Foucault and Political Reason: Liberalism, Neo-liberalism and Rationalities of Government*. Chicago: University of Chicago Press.
Benfell, P. (2003) 'Science grants all about sustaining development', *New Zealand Herald*, 12 September: A13.
Blythman, J. (2003) 'The dark truth about UK supermarkets', *Grower*, 58 (3): April: 22–5.
Bonanno, A., Busch, L., Friedland, W., Gouveia, L. and Mingione, E. (eds) (1994) *From Columbus to ConAgra: The Globalization of Agriculture and Food*. Lawrence (KS): University of Kansas Press.
Bourne, R. (2003) 'United front', *Horticulture News*, July: 26–30.
Bull, P. (2003) 'Fruitful pickings in added value', *New Zealand Herald*, 8 September: C4.
Caradus, J. (2003) 'Exciting early era at Dexcel', *Dairy Exporter*, 79 (2): September: 64.
Dann, L. (2003a) 'Zespri conquers the odds', *New Zealand Herald*, 18 July: C1.
Dann, L. (2003b) 'Land being squeezed on research', *New Zealand Herald*, 15 September: C4.

Dean, M. (1999) *Governmentality: Power and Rule in Modern Society*. London: Sage.

Editorial (2003) 'Soil sciences funding cut harms nation', *New Zealand Herald*, 5 September: A14.

Fairweather, J. and Campbell, H. (2003) 'Environmental beliefs and farm practices of New Zealand farmers: opposing pathways to sustainability', *Agriculture and Human Values*, 20 (3): 287–300.

Foundation for Research, Science and Technology (FoRST) (2002) *Investment Signals and Request for Proposals, Sustainable Development Portfolios, for Research in Sector Sustainability Portfolios, Sustainable Management of Productive Sector Environments and Sustainable Cities and Settlements*. Wellington: FoRST, October.

Freeth, A. (2003) 'Research lost its way', *New Zealand Herald*, 22 September: E10.

Friedland, W. (2001) 'Commodity systems analysis – a reprise', *International Journal of the Sociology of Agriculture and Food*, 9 (1): 82–103.

Hartley, A. (2003) 'Genetic modification – what the government is doing', *Newsletter*, Ann Hartley, MP for Northcote, September.

Hayward, D. and Le Heron, R. (2002) 'Horticultural reform in the European Union and New Zealand: further developments towards a global fresh fruit and vegetable complex', *Australian Geographer*, 33 (1): 9–27.

HortResearch (2000) *Annual Report*. Palmerston North: HortResearch.

Jessop, B. (1999) 'Reflections on the (il)logics of globalisation'. In K. Olds, P. Dicken, P. Kelly, L. Kong and H. Yeung (eds) *Globalisation and the Asia Pacific: Contested Territories*. London: Routledge, pp. 19–38.

Jessop, B. (2000) 'The state and the contradictions of the knowledge-driven economy'. In J. Bryson, P. Daniels, N. Henry and J. Pollard (eds) *Knowledge, Space and Economy*. London: Routledge, pp. 63–78.

Johnson, R. (1992) 'Agricultural debt'. In S. Britton, R. Le Heron and E. Pawson (eds) *Changing Places in New Zealand: A Geography of Restructuring*. Christchurch: New Zealand Geographical Society: 97–101.

Khachatourians, G. (2001) 'How well understood is the "science" of food safety'. In P. Phillips and R. Wolfe (eds) *Governing Food: Science, Safety and Trade*. Kingston: McGill-Queen's University Press, pp. 14–23.

Kloppenburg, J. (1988) *The First Seed: The Political Economy of Plant Biotechnology*. Cambridge: Cambridge University Press.

Larner, W. and Le Heron, R. (2004) 'Global benchmarking: participating at a distance in the global economy'. In W. Walters and W. Larner (eds) *Global Governmentality*. London: Routledge, pp. 212–32.

Le Heron, R. (1989a) 'A political economy perspective on the expansion of New Zealand livestock farming, 1960–1984. Part I: Agricultural policy', *Journal of Rural Studies*, 5 (1): 17–32.

Le Heron, R. (1989b) 'A political economy perspective on the expansion of New Zealand livestock farming, 1960–1984. Part II: Farmer Responses – Aggregate Evidence and implications', *Journal of Rural Studies*, 5 (1): 33–41.

Le Heron, R. (2003) 'Cr(eat)ing food futures: reflections on food governance issues in New Zealand's agri-food sector', *Journal of Rural Studies*, 19 (1): 111–125.

Le Heron, R. and Roche, M. (1996a) 'Eco-commodity systems: historical geographies of context, articulation, and embeddedness under capitalism'. In D. Burch, R.E. Rickson and G. Lawrence (eds) *Globalisation and Agro-food Restructuring: Perspectives in the Australasian Region*. Aldershot: Avebury, pp. 73–89.

Le Heron, R. and Roche, M. (1996b) 'Globalisation, sustainability and apple orcharding, Hawkes Bay, New Zealand', *Economic Geography*, 72 (4): 416–32.

Le Heron, R. and Roche, M. (1999) 'Rapid reregulation, agricultural restructuring and the reimaging of agriculture in New Zealand', *Rural Sociology*, 64 (2): 203–218.

Lynch, B. (2001) Interview: Executive Director of the Meat Industry Association, July, Wellington.

Lynch, B. (2003) Interview: Executive Director of the Meat Industry Association, August, Wellington.

MacLeod, G. (2001) 'New regionalism reconsidered: globalization and the remaking of political economic space', *International Journal of Urban and Regional Research*, 25 (4): 804–29.

MacLeod G. and Goodwin, M. (1999) 'Space, scale and state strategy: rethinking urban and regional governance', *Progress in Human Geography*, 23: 503–27.

Manhire, J., Campbell, H. and Fairweather, J. (2002) 'Draft plan for research: pathways for sustainability. Comparing production systems across four sectors of New Zealand agriculture', *Centre for the Study of Agriculture, Food and Environment {CSAFE} Discussion Paper 2.*

McKenna, M. and Murray, W. (2002) 'Jungle law in the orchard: comparing globalization in the New Zealand and Chilean apple industries', *Economic Geography*, 78 (4): 495–514.

Meat Industry New Zealand (2002) *Annual Report*. Wellington: Meat Industry New Zealand.

Meat Producers Board (1995) *Annual Report*. Wellington: Meat Producers Board.

Meat Producers Board (1999) *Annual Report*. Wellington: Meat Producers Board.

Rose, N. (1999) *Powers of Freedom: Reframing Political Thought*. Cambridge: Cambridge University Press.

Sandrey, R. and Reynolds, R. (1991) *Farming Without Subsidies*. Wellington: GP Publishers.

Tanaka, K., Juska, A. and Busch, L. (1999) 'Globalization of agricultural production and research: the case of the rapeseed sector', *Sociologia Ruralis*, 39 (1): 54–77.

Tyson, J. (2003) 'Mark Jefferies: Managing the change', *Meat Producer*, 31 (2): 33–4.

5 Contesting biotechnology
Cross-continental concerns about genetically modified crops

Yolanda Massieu and Michelle Chauvet

Introduction

Recent analyses of cross-continental food systems have emphasized processes of *re-regulation*, in which trade flows are increasingly circumscribed by an array of regulatory issues and documentary requirements for traceability.[1] These processes have also been apparent with regard to genetically modified (GM) crops, but as this chapter illustrates, issues relating to traceability have taken a complicated and contested path.

Genetic modification (GM) advocates and apologists argue that these products are generally safe for the environment and for human health, and so these new regulatory arrangements and governance structures represent an unnecessary burden on trade. But in any case, how effective are current restrictions on the trade and cultivation of GM crops? The purpose of this chapter is to examine the background to contemporary contestation over this issue, and to assess contemporary attempts to regulate the sector, specifically concerning the efficacy of the Cartagena Biosafety Protocol. It argues that current arrangements do not provide a comprehensive basis for national governments to restrict GM crops. The Cartagena Biosafety Protocol, a multilateral agreement that seeks to define the terms by which national governments can place restrictions on GM crops, is revealed as having substantial limitations. Furthermore, even when governments impose limitations on the GM sector, there exist practical difficulties in realizing these goals. Referring to the case of Mexico and GM maize, the chapter notes that although the Mexican government imposed a moratorium on GM maize cultivation, the country's deepening dependence on (subsidized) US agricultural exports has exposed the country's consumers to GM maize, and has led to genetic contamination.

Biotechnology, agricultural markets and consumers

The development of GM crops has given rise to much widespread and heated debate on the possible consequences – both positive and negative – of this new technology. In 2002 it was estimated that GM crops were grown on

58.7 million hectares by 5.5 million farmers in 16 countries. Some 99 per cent of this production was concentrated in four countries: the US, Canada, Argentina and China. The principal GM crop is soybean, occupying 36.5 million hectares (62 per cent of the total global area of GM crops), followed by GM corn at 12.4 million hectares (21 per cent of the total), GM cotton at 6.8 million hectares (12 per cent of the total) and GM canola at 3 million hectares (5 per cent of the total). The cultivation of herbicide tolerant plants has been the dominant rationale for the adoption of GM varieties (accounting for 75 per cent of GM crops planted between 1996 and 2002), while insect resistance varieties accounted for 17 per cent of the global crop, and stacked genes accounted for 8 per cent (James 2002).

The supporters of GM crops argue that there are minimal risks with the consumption of GM foods, because at the DNA level all organisms are the same. Nature works with bacteria and viruses which continuously change their DNA structures. As such, deliberate genetic modification by scientists is argued to represent an adaptation of what essentially remain natural processes. Further, biotechnology firms defend these technologies because of their allegedly vital importance in securing global food production for a growing world population. Recent marketing activities by many of these companies make the argument that famine and poverty are more likely without widespread adoption of GM crops.

Both these arguments deserve scrutiny. The issues of whether genetic modification is a natural or artificial process and whether it is safe or not raise complex ethical and scientific questions. The issue of GM crops and world hunger is equally debatable. A recent report by the Food and Agriculture Organization (FAO 2003) argues that world food demand is being reduced and, in any case, increased levels of agricultural production do not necessarily resolve problems of hunger.[2] Smallholder farmers in developing countries generally cannot afford GM seeds, so these technologies may not assist food security for these groups. They are most appropriate for large-scale agriculture, which tends to be owned and controlled by agribusiness interests. In this sense, it has been argued that biotechnology may be contributing to an increase in the problem of food insecurity, not its resolution (Walsh 2000). In addition, the main new traits of GM seed are aimed at herbicide resistance in commercial crops (mainly soybean), rather than addressing the specific agronomic problems of local crops vital to rural householder subsistence. Accordingly, it can be argued that the use of arguments about world hunger by biotechnology firms represents a simplistic and illegitimate incursion into this debate, which serves to justify the further expansion of GM crop interests by the North's transnational corporations.

A further set of issues with regard to GM crops is raised by changing consumer attitudes towards these products. Since the 1980s there has been an increasing interest from civil society about food and plant genetic resources; this is no longer an exclusive domain of specialized scientists. Genetic resources have now a more important international dimension since the conflict occurs as part of North–South political and economic relations (Pistorius and van Wijk 1999: 7). Consumer attitudes towards GM products are increasingly affecting the shape of agriculture trade. As summarized by Wilkinson (2003: 24):

> When transgenics came to market in the middle '90s they encountered a very different agrofood system than that which had prevailed in the early heady days of biotechnology research. A fundamental shift from price to quality criteria in food consumption had taken place.

Public debate concerning the consequences of modifying nature has a high and important profile in many countries, and in response, leading food retailers have been required to adapt their commercial practices. For example, in the late 1990s British supermarkets very publicly announced their intentions of not selling GM foods (Walsh 2000). More recently, the world's two largest food companies, Nestlé and Unilever, have also declared publicly that they will not sell transgenic food. Major food companies have refused to use genetically modified potatoes in their processed potato products for fear of consumer backlash, and this has resulted in decreased production of GM potatoes (Curtis *et al.* 2003). Seen in its broader terms, this is part of a wider consumer trend, which is favouring organics and fair trade food.

The attempted resolution of international contestation over GM crops through the Cartagena Biosafety Protocol

During recent years the EU and US have adopted widely different positions with regard to the regulation of GM foods. The US has strongly advocated their use, whereas the EU has taken a cautionary stance.[3] This has led to prolonged dispute between these parties, one manifestation of which has been a complaint put by the US to the WTO over the EU's position on GM foods.

Central to this dispute is the question of how to evaluate the costs, benefits and risks of GM crops. The EU has taken the view that extensive public dialogue is necessary for the implementation of best practices in decision-making about GM foods (Shenkelaars 2001; Commission of the European Community 2002). Although EU members have different visions and opinions about this subject (Commandeur *et al.* 1996), European risk evaluation procedures generally involve a relatively extensive frame of reference, in which the broad-ranging assessment of socio-economic and environmental impacts is included within cost-benefit assessment. In contrast, US procedures since the Reagan era have tended to be more narrowly focused around a range of economic and technical considerations (König 2002). Governing the US approach to this issue are the operations of the Agricultural Bio-technology Research Advisory Committee (ABRAC), a consultative agency of the US Department of Agriculture (USDA), established in 1986.

A critical element of the US approach is its opposition to the use of the precautionary principle as a binding principle of international trade law when assessing the implications of GM crops. Successive US administrations have been against any inclusion of the precautionary principle in these contexts, and have not accepted the inclusion of wide-ranging social and environmental assessments in risk evaluation. In general, these are viewed as protective barriers to free commerce in GM products. Of course, this position sits well with the country's economic interests, given that the US is a major exporter of GM crops. Nevertheless, both EU and US corporations hold most of the intellectual property relating to this sector, which means that the EU also has incentives to participate in biotechnology, despite its different policies to the US.

US opposition to the inclusion of the precautionary principle in inter-national law relating to this sector was manifested most clearly in the drafting of the Cartagena Biosafety Protocol. This agreement is a component of the Convention of Biological Diversity, and aims to specify the conditions under which countries can regulate the flow of GM products. The negotiation of the protocol took five years, after which it was finally signed in 2000 by 130 countries (Luna 2000: 52–5).

The protocol's final form expresses a consensus among the widely diver-gent views towards trade in genetically modified organisms (GMOs). On the one hand, it acknowledges international concerns over the expansion of biotechnology with regard for biological diversity and human health. Specifically, it endorses the need for special care in biological centres of origin (see p. 71), and it approves the use of the precautionary principle by coun-tries when assessing the risks, costs and benefits of GM products. This means that a country can restrict the entry of GM products when there are doubts about possible harm to the environment and/or public health (United Nations Convention on Biological Diversity 2000: 1). On the other hand, the protocol is established as international 'soft law', which is subservient to other inter-national agreements, notably those relating to trade that are bound within the structures of the World Trade Organization. Consequently, its cautionary spirit is contradicted by its requirement to ensure the maintenance of commerce along the broadly neo-liberal lines of WTO agreements.

Furthermore, the subservience of the Cartagena Biosafety Protocol to WTO agreements ensures a view of genetic resources as private property. The TRIPS agreement (Trade Related Aspects of Intellectual Property Rights), part of the WTO family of trade agreements, allows patents on living organisms. Of course, this is highly controversial because, strictly speaking, living organ-isms are not an invention. Moreover, this view conflicts with the holistic traditions of many cultures, especially those of indigenous people, for whom life cannot be rendered as private property. Nevertheless, the possibility of patents over agricultural genetic resources is now a component (albeit a contested one) of international law, and this further complicates the governance structures of cross-continental food systems.

This status of the Cartagena Biosafety Protocol vis-à-vis other international agreements strongly reflects the bargaining position of the US. During the drafting of the protocol, the US demanded the incorporation of wording that stipulated explicitly that the inclusion of the precautionary principle did not exempt signatory countries from their obligations to international commitments under the World Trade Organization (König 2002).

Although the EU and US both hold the common position that decisions about GM crops should be taken on the basis of scientific knowledge and risk assessment, they differ in their views of how this should operate in practice. Specifically, views differ with regard to: the operation of the precautionary principle; the substantial equivalence criterion for GMO authorization; the labelling and segregation of exports; and, finally, the scope of risk assessment (with the US not recognizing the inclusion of social and environmental impacts in the evaluation and handling from the risk).[4]

Contestation over these issues has a range of implications for the governance structures of international food systems. First, they raise issues for the international trade of processed food products making use of GM ingredients. In general, the Cartagena Protocol restricts countries from barring processed foods that make use of GM ingredients, on the assumption that GMOs pose no environmental harm when processed. However, strong consumer resistance to GM foods in some countries (mainly in Europe) has led a number of supermarket chains and some processing companies to publicly state they are 'GM free'. In order to comply with this assertion, these entities require suppliers to maintain traceability systems and to segregate GM from non-GM products.

The situation is different for unprocessed products – notably crops such as soybean and maize – that originate from GM seed stock. Under the Cartagena Protocol, countries are entitled to restrict these imports if there are grounds to reason that they may impact negatively on environmental or public health. However, this may be easier said than done, because GM crops are not necessarily identified as such.

Supporters of GM products tend to argue that the Cartagena Protocol reflects a new type of agricultural trade barrier. However, as outlined in the following discussion of biodiversity, biotechnology and GM contamination in Mexico, these issues have considerably greater complexity, and relate to the interaction of economics, social processes and the environment.

Mexico and transgenic maize

Recent events in Mexico with regard to transgenic maize bring into sharp focus the range of issues attached to debates on the regulation of GM products. Moreover, this is an internationally important case because of its implications for biodiversity and food security in a developing country context. As will be explained below, Mexico is the genetic centre of origin of maize, and the crop is a vital staple for the nation's population. Accordingly, there are high stakes in the debate over the entry of transgenic maize into Mexico.

In ecological terms, this issue highlights the global geographical divide between the countries best able to exploit commercially genetic resources and those that are the sites for most genetic diversity. The process to obtain a commercial plant variety from a wild one needs years of research and investment. This means that only those countries that dedicate enough funds to agriculture research are able to decide and exploit plant genetic resources. These conditions are found mainly in industrialized countries. However, the bulk of the world's genetic diversity is located in a group of developing nations, known collectively as the Vavilov centres of biological diversity (Table 5.1). Moreover, these are also the places where, because of biodiversity, potential environmental risks arising from GM contamination are greatest. As noted by Rissler and Mellon (1996: 22): 'Genetically engineered crops are not inherently dangerous; they only present problems where the new traits . . . produce unwanted effects on the environment.' The main environmental risks are related to the possibility of genetic crosses with non-transgenic crops, leading to the appearance of new weeds, plagues and/or the disappearance of landscapes' important crops. Additionally, concerns have been aired with regard to health, allergies and toxicity.

These issues are vitally important in consideration of Mexico, one of the world's most important Vavilov centres. Mexico is the centre of origin for maize, a vital food crop in feeding the human race (Mooney 1979). In Mexico, agricultural genetic diversity has been reduced progressively for a number of decades, because of the effects of Green Revolution hybrids. Globally, food crops' genetic diversity diminished 75 per cent during the twentieth

Table 5.1 Vavilov centres of biological diversity

Region	Origin crops
Central America	Maize, tubercles
Andes	Potatoes, peanuts
South Brazil, Paraguay	Manioc
Mediterranean	Oats, canola
South-west Asia	Rye, barley, wheat, green pea
Ethiopia	Barley, sorghum, millet
Central Asia	Wheat
Indo-Burma	Rice, dwarf wheat
South-East Asia	Banana, sugar cane, yam, rice
China	Fox tail millet, soybean, rice

Source: Vélez and Rojas (1998).

century, and in Mexico, only 20 per cent of the agricultural varieties grown in the country in 1930 were still being cultivated by the end of the century (GRAIN 1996).

The reduction of genetic diversity during recent decades and the potential for this to be accelerated through the introduction of GM crops present important repercussions for small subsistence peasant production in Mexico. In particular, the prospects of genetic contamination could have serious implications for the cultivation of maize within the complex inter-planting practices of small peasant landholders. In Mexico, maize is a staple food linked to cultural identity. It is the basis of subsistence agricultural production, supplying tortilla and other foods for the family, as well as feeding livestock. In the State of Chiapas, for example, peasants are considered to be low yield cultivators of corn (their maize yields are just two tonnes per hectare), but this fails to take into account the fact that this crop is inter-planted with beans, squash, vegetables and fruits, and taken together, these landholders generate total food yields of 20 tonnes per hectare (Shiva 2000: 4).

In the opinion of many non-governmental organizations (NGOs), these cultivation practices are imperilled by the introduction of GM maize. The major concern rests with the potential for contamination of the existing crop, altering the genetic profiles and characteristics of traditional maize varieties. For these reasons, in 1999 the Mexican government implemented a moratorium on cultivation of GM maize, even for field trials. Nevertheless, despite these attempted restrictions, Mexico's attempt to remain 'GM-free' was soon breached.

The reality of the threat of genetic contamination was brought into sharp focus in August 2000 via the 'StarLink' case. In this case, GM corn that was forbidden for human consumption found its way into Taco Bell and Kraft Food products. Moreover, this contamination was revealed not by any regulatory authority but by the NGO 'Friends of the Earth'. As a consequence of this revelation, Kraft Foods had to retire 300 products from the market in September 2000, while the owner of StarLink corn, Aventis, stopped Star-Link seeds sales, and the USDA retired 350,000 acres planted with this transgenic corn (López Villar 2003). Following the exposure of this contamination, StarLink corn was also found in US corn exports to Japan and Korea. This case underscores the difficulties of containing and controlling the GMO presence, once a particular form has been approved for the market.

A second episode of GM contamination was then exposed in 2001, when two researchers from the University of California at Berkeley (Dr Ignacio Chapela and Dr David Quist) published evidence of GM maize in crop samples from the north of Oaxaca. The publication of these results in the journal *Nature* generated a major scandal in Mexico. The National Ecology Institute and the National Biodiversity Commission commissioned two of the country's leading research institutions (Universidad Nacional Autónoma de México (UNAM) and Centro de Investigación y Estudios Avanzados (CINVESTAV)) to undertake further studies into these allegations. These studies confirmed Chapela and Quist's findings, although *Nature*, after publishing their results, expressed some doubts. At the time of writing, the results of a further study into this issue have not been published, and the issue remains contested. Nevertheless, the Director of the National Ecology Institute has said GM contamination is present not only in Oaxaca, but also in the State of Puebla. Most recently (in 2004), however, Dr Amanda Gálvez, president of the Mexican government's Inter-ministries Biosafety Commission Consultative Council, and Dr Ariel Alvarez, a Mexican researcher, declared to the media that transgenic pollution is minimal in maize cultivars, as only 7.6 per cent of 200 plots studied showed evidence of GM varieties. Alvarez stated that he assumed this is because transgenic maize is not as productive as local varieties, leading to few peasants planting these varieties (*La Jornada* 2004).

The breaches in Mexico's moratorium on GM maize represent a serious problem with the international regulation of GM food trade. Because of the lack of appropriate monitoring and separation systems, transgenic maize varieties are now spreading in the crop's centre of origin, with unknown consequences. On the one hand, this damages the ability of Mexican producers to market their maize as 'GM-free'. On the other, it potentially affects the genetic qualities of maize varieties used by smallholder agriculturists, denying farmers' rights to select and use non-transgenic seeds. Until now, Mexican authorities have not developed an effective response to these problems, although environmental and peasant organizations have demanded an end to GM maize imports from the US, the most probable source of pollution. A group of these organizations has requested intervention from the North American Commission for Environmental Cooperation (CEC), a tri-national commission established in the NAFTA context, and at the time of writing, a report is expected to be published in the second half of 2004.

The evidence of GM maize contamination clearly illustrates regulation problems in Mexico towards biotechnology. There appears to be a lack of government interest to protect both basic food production and maize's genetic diversity, and there are severe contradictions in government institutions. On the one hand, there exists an inter-departmental commission to regulate transgenic crops planting in Mexico that forbids the use of transgenic maize in the country; whereas on the other hand the Ministry of Economics allows transgenic maize to be imported into Mexico for consumption. In turn, Mexico's dependence on maize imports is a consequence of economic policies that have neglected internal maize production for decades (Massieu and Lechuga 2002), in the contexts of significant agricultural subsidization by the US and the NAFTA. In contradiction to Mexico's commitments under the Cartagena Biosafety Protocol, a biosafety law for the country has not been drafted.

Social movements and contestation over transgenic crops in Mexico

Opposition to the introduction of GM maize in Mexico has brought together a wide range of social actors representing environmental, cultural and health interests, which at national level have forged alliances between with peasant movements.[5] The most notable of these are joint actions by Greenpeace-Mexico and the Unión Nacional de Organizaciones Regionales Campesinas Autónomas (UNORCA), a leading Mexican peasant organization which is a member of Via Campesina, the global NGO representing peasant agriculturists. Both Via Campesina and UNORCA consider the introduction of GM seeds as a threat to peasant communities, as it represents a loss of autonomy and an increased economic and technological dependence on transnational corporations (Poitras in press). The cooperation of these two organizations represents the most recent chapter of Greenpeace-Mexico's long-running campaign against GM products, which began in 1998.

Greenpeace-Mexico operates according to a different model compared with other Mexican civil society organizations. It is styled on social movements found in industrialized countries, and is not linked organically with the political history of the country. Greenpeace-Mexico has no broad popular constituency or large membership, nor does it get much financing from local sources, relying instead largely on payments from Greenpeace International (Covantes 1999). It makes extensive use of the media, focuses on non-traditional issues, and relies heavily on its transnational network. This contrasts considerably with mass membership Mexican peasant and indigenous movements, which correspond more closely to traditional models of political agency and collective action.

Greenpeace has drawn considerably on the symbolic power of maize in Mexico to make its case, noting not only the ecological dangers from GM contamination, but also the hardship suffered by national producers confronted with increasing amounts of subsidized maize from the US. Moreover, through its numerous actions around the issue, Greenpeace taps into a still very resonant nationalist and anti-imperialist chord among the Mexican population.[6] In order to prove that transgenic maize was being imported into Mexico – something the government was denying – Greenpeace contracted a laboratory from Austria to test samplings of maize imported from the US, and then went to the Ministry of Agriculture (SAGAR) with the evidence. SAGAR passed the claim to the Ministry of Health, but the latter declined to deal with the issue, saying that it was SAGAR's responsibility. Greenpeace thus successfully exposed the contradictions and gaps in the regulation of the introduction of transgenic crops in the country and argued for the opening up of biotechnology regulation and monitoring institutions to civil society organizations.[7]

Conclusions

The intention of this chapter has been to discuss the conflicting processes that lie at the heart of the international regulation of GM crops. On the one hand, through the Cartagena Biosafety Protocol national governments have sought to develop a multilateral framework to govern the trade and cultivation of these products. However, as discussed above, this regulatory device is generally weak. Because it prioritizes commerce over the precautionary principle, it gives impetus for the further expansion of the GM sector. On the other hand, however, regardless of these initiatives, governments may face considerable practical difficulties in regulating these sectors. As illustrated through the example of transgenic maize in Mexico, mandated restrictions by the Mexican government have not necessarily preserved the country as 'GM-free'. Given the significance of Mexico as the 'centre of origin' for maize, these developments potentially hold important repercussions for national and international biodiversity and food security.

It is clear that agro-biotechnology is transforming cross-continental food systems and agriculture trade. The status of transgenic products trade is a major theme in contemporary international trade negotiations, and in the private sector, supermarket firms and traders are increasingly placing traceability requirements on upstream producers. Political contestation over these issues is apparent at a range of scales, and social movements are becoming increasingly influential in shaping corporate actions and government policies. Accordingly, with regard to transgenic products trade the global agri-food system is at a special historical moment, and the resolution of issues currently in contestation will have wide influence over the future shape of the global food economy.

Notes

1 Traceability means the information and control system about the 'trail' of foodstuffs. It allows access to full information about the product and leads to better guarantees of food safety.

2 World population will grow from around 6 billion people today to 8.3 billion people in 2030. Population growth will be growing at an average of 1.1 per cent a year up to 2030, compared to 1.7 per cent annually over the past 30 years. At the same time, an ever increasing share of the world's population is well fed. As a result, the growth in world demand for agricultural products is expected to slow further, from an average 2.2 per cent annually over the past 30 years to 1.5 per cent per year until 2030. In developing countries, the slowdown will be more dramatic, from 3.7 per cent for the past 30 years to an average of 2 per cent until 2030 (FAO 2003).

3 When talking about the EU it is necessary to note that responsibilities for these issues are divided between the central institutional apparatus of the Union and the constituent members' national legislatures. The enactment into law of common rules regarding GM foods requires passage of legislation through national parliaments, a process that can take up to two years (Schenkelaars 2001).

4 '... the industry claims that there is "substantial equivalence" between genetically engineered products and natural ones. When corporations claim monopoly rights to

seeds and crops, they refer to GMO as "novel". When the same corporations want to disown risks by stifling safety assessment and analysis of hazards, they refer to transgenic organisms as being substantially equivalent to their naturally occurring counterparts. The same organisms cannot be both "novel" and "not novel". This ontological schizophrenia is a convenient construct to create a regime of absolute rights and absolute irresponsibility. Through the WTO the ontological schizophrenia is being spread from the United States to the rest of the world' (Shiva 2000).

5   Although there is no evidence that the consumption of GM maize is harmful to human health, its transgenic qualities are not eliminated through processing, and in this respect some consumers remain wary of consuming this product.

6   For instance, during 1999, both the Angel del Paseo de la Reforma, a monument to the heroes of the Independence in Mexico City, and a historic fortress in Veracruz, also symbolizing Mexican resistance to imperialism, have been occupied by Greenpeace activists, who announced the 'Mexican declaration of genetic independence'. Moreover, a huge banner was displayed denouncing the US imperialism underpinning the import of transgenic maize to Mexico. On 12 September 2003 in Veracruz, they stopped a ship with corn imports.

7   The sources for this section on Greenpeace include Mexican press coverage of their actions, an interview with Liza Covantes, head of the Genetic Engineering Campaign at Greenpeace-Mexico in Mexico City (Covantes 1999), and various press releases and documents produced by Greenpeace.

References

Commandeur, P., Joly, P.B., Levidow, L., Tappeser, B. and Terragni, F. (1996) 'Public debate and regulation of biotechnology in Europe', *Biotechnology and Development Monitor*, 26: 2–9.

Commission of the European Community (2002) 'Life sciences and biotechnology – a strategy for Europe', *COM*, 27 (final), 23 /01, Brussels.

Covantes, L. (1999) Personal communication, Mexico City, 13 April.

Curtis, K., McCluskey, J. and Wahl, T. (2003) 'Is China the market for genetically modified potatoes?', *AgBioForum*, 5 (4): 175–8.

Food and Agriculture Organization (FAO) (2003) *World Agriculture: Towards 2015/2030.* Rome, FAO.

Genetic Resources Action International (GRAIN) (1996) 'The biotech battle over the golden crop', *Seedling* 13 (3), in Greenpeace (2000) *Centros de Diversidad*, Greenpeace: Mexico City: 5.

James, C. (2002) 'Global status of commercialised transgenic crops: 1999', *ISAA Briefs*, 12, Ithaca, New York.

König, A. (2002) 'Negotiating the precautionary principle: regulatory and institutional roots of divergent US and EU positions', *International Journal of Biotechnology*, 4 (1): 61–79.

La Jornada (2004) 'Centíficos desestiman riesgos potenciales para las personas, la agricultura y el ambiente', Sec. Sociedad y Justicia (14 February): Mexico City.

López Villar, J. (2003) *Contaminación Genética*. Amsterdam, Friends of the Earth International GMO Program.

Luna, D. (2000) 'Protocolo sobre la seguridad de la biotecnología del Convenio sobre la Diversidad Biológica', *CONACOFI Annual Meeting Proceedings* (24–26 October): Mexico City: 52–5.

Massieu, Y. and Lechuga, J. (2002) 'El maíz en México: biodiversidad y cambios en el consumo', *Análisis Económico*, 36, Universidad Autónoma Metropolitana-Azcapotzalco, Economy Department, Mexico.

Mooney, P.R. (1979) *Seeds of the Earth. A Private or Public Resource?* Ottawa and London, Canadian Council for International Cooperation and the International Coalition for Development Action.

Pistorius, R. and van Wijk, J. (1999) *The Exploitation of Plant Genetic Information. Political strategies in crop development*, Biotechnology and Agriculture Series. Cambridge: CABI Publishing, University Press.

Poitras, M. (in press) 'Social movements and techno-governance: reclaiming the genetic commons. The case of genetic engineering in Mexico'. In G. Otero and M. Poitras (eds) *Food for the Few: Agrobiotechnology and Global Neoliberalism in Latin America*. London: Zed Books.

Rissler, J. and Mellon, M. (1996) *The Ecological Risks of Engineered Crops*. Boston (MA): Institute of Technology.

Shenkelaars, P. (2001) 'Uncertainty and reluctance: Europe and GM foods', *Biotechnology and Development Monitor*, 47 (9): 16–19.

Shiva,V. (2000) *Stolen Harvest: The Hijacking of the Global Food Supply*. Cambridge (MA): South End Press.

United Nations Convention on Biological Diversity (2000) *Cartagena Biosafety Protocol*. Online. www.biodiversidadla.org/documentos3.5.htm (accessed 10 December 2000).

Vélez, G. and Rojas, M. (1998) 'Definiciones y conceptos básicos sobre Biodiversidad', *Biodiversidad, Sustento y Culturas*, Cuadernillo 1, Programa Semillas, Bogotá, Colombia.

Walsh, V. (2000) 'Creating markets for biotechnology', unpublished paper presented at the *International Rural Sociological Association X World Congress*, Rio de Janeiro, Brazil.

Wilkinson, J. (2003) 'Biotechnology, the agrofood system: science and the consumer', unpublished paper presented at the *Conference Cycle La Agricultura Transgénica: Pros y Contras ¿para Quién?* of the Asociación Mexicana de Estudios Rurales, Sociology Department of Universidad Autónoma Metropolitana-Azcapotzalco and CONACYT.

Part II

The local impacts of cross-continental food chains

6 The cross-Pacific chicken

Tourism, migration and chicken consumption in the Cook Islands

Jane Dixon and Christina Jamieson

Introduction

Appadurai (1990) contends that globalization is constructed as much by flows of labour, technology, ideas and media as it is by flows of finance capital and commodities. This is amply evident in relation to food, where globalization has been associated with the disembedding of diets from tradition and local geographies (Pelto and Pelto 1983; Probyn 1998; Dixon 2002). Typically, these processes are said to destroy unique culinary cultures by subsuming local customs and distinctive exchange relations to a process of 'Coca-colonization', or the spread of Western consumption practices (Howes 1996). Yet, such accounts tend to imagine a static and one-dimensional set of interactions between globalization and national food systems. Anthropologists, in particular, reveal that local communities are more than capable of customizing commodities in the act of consumption (James 1996; Howes 1996; Miller 1997). Moreover, it is difficult to generalize on the balance of economic and social benefits and costs from global connectivity in the food system (Alexeyeff 2004). In *Cosmologies of Capitalism*, Sahlins (1994: 415) refers to the ways in which peoples in the Pacific Islands have incorporated Western goods and persons into their own ceremonial exchanges, social valuables or sacred customs and suggests that 'the exploitation by the world system may well be an enrichment of the local system'.

In these contexts, this chapter examines the impact of tourism (an incoming flow) and migration (an outgoing flow), as two manifestations of globalization, on the culinary culture of a small Pacific island state, the Cook Islands. With particular attention to changes in the cultural economy of chicken meat, it reveals the complex and apparent contradictory dynamics that have led Cook Islanders to incorporate imported product into their culinary repertoires while at the same time shunning local birds. More specifically, tourists to the Cook Islands demand 'authentic island' experiences that include palm-fringed beaches, a slow pace of life and, most pertinently to this chapter, banquets featuring island-style foods. On the one hand, this provides an arena for Cook Islanders to continue to maintain and perform traditional food-based rituals, albeit in the decontextualized settings of modern hotels. Yet at the same time, the only way that these hotels can provision large numbers of tourists with what they imagine to be an 'authentic island' food experience is to rely on sizeable imports of chicken meat. Therefore, while tourism is encouraging the reproduction of the feasting aspect of the Cooks Islands' culinary culture, this is achieved on the basis of an internationalized food economy. Furthermore and in a similar vein, the constant travels by Cook Islanders between their homeland and cosmopolitan centres is leading to significant modifications in other parts of their culinary culture. Both developments rely on imported foods and ideas, including those about convenience and the role of food in social life. However, neither development has buttressed the sustained development of local agricultural production.

Background to the Cook Islands

The Cook Islands are an isolated Polynesian nation comprising 15 islands (11 of which are inhabited) spread over 2.2 million square kilometres of the South Pacific Ocean (Figure 6.1). They were under New Zealand administration until 1965, when they became internally self-governing with New Zealand retaining responsibility for foreign policy and defence. This relationship of 'free association' provides Cook Islanders with New Zealand citizenship and the right to enter at will both New Zealand and Australia. In 2001 the population was recorded as 18,027, which was a 5.6 per cent decrease from 1996. While the capital (Rarotonga) grew by almost 8 per cent, the other islands experienced a decrease in population of 26 per cent.

As Polynesians, Cook Islanders see themselves as travelling peoples. A history of ocean voyaging and settling new islands is recalled in genealogies, songs, dances and legends. This deep sense of travelling history and identity now sits alongside more recent experiences. Following the Second World War there was a diasporic movement of Islanders to Pacific Rim metropoles. The 1974 opening of the Rarotonga international airport is remembered as much for Cook Islanders' departing as for foreigners arriving (Jamieson 2002: 70–3). By the 1970s, the population of Cook Islanders resident in New Zealand exceeded that of the Cook Islands itself. Now, more than 85 per cent of Cook Islanders live outside their homeland, with approximately 60,000 in New Zealand and sizeable numbers also in Australia and the US. In addition to these migratory flows, Cook Islanders also embark frequently on trips of shorter duration, often in church or dance groups. These trips (known as *tere* parties – '*tere*' meaning 'to trip or voyage') typically involve fundraising, staying with relatives, shopping for clothing and other goods not readily available at home, sightseeing and social activities such as feasting and dancing. Thus Cook Islanders, though ocean voyagers for millennia, have also become tourists in the sense of their engagement in the commodified and globalized tourist industry (Jamieson 2002: 78).

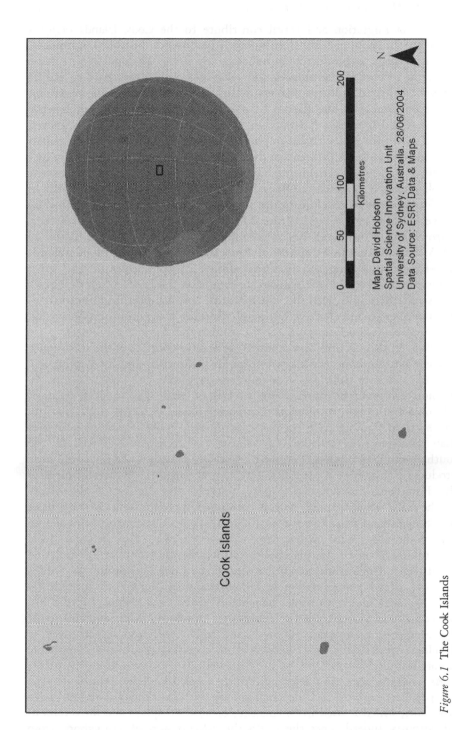

Figure 6.1 The Cook Islands

Source: David Hobson, Spatial Science Innovation Unit, University of Sydney, 28 June 2004, sourced from ESRI Data & Maps.

Flows of migration and travel contribute to the Cook Islands being a 'MIRAB' society – one whose economy is based on *mi*gration, *r*emittances, *a*id and a large, unproductive *b*ureaucracy (Syed and Mataio 1993). At one level, the receipt of remittances and relatively high wages paid to public servants and pearl industry workers means that the Cook Islands are among the most affluent of the Pacific Islands, with comparatively high literacy levels (World Bank 2002: 10). These indicators, however, mask an uneven spread of wealth and extreme import dependence, particularly with regard to food. In 1982, 25 per cent of the Cook Islands' gross domestic product (GDP) originated in agriculture, but by 2000 it had declined to 15 per cent (Cook Islands (CI) Statistics Office 2002). Subsistence agriculture continues to be important, with close to one in two households being agriculturally active (CI Ministry of Agriculture 2001: 17). According to one researcher, '"growing your own food" has a near religious imperative' (Syed and Mataio 1993: 71). With the exception of Rarotonga, 75 per cent of households are engaged in fishing activities and almost all with access to land grow some taro and coconut for regular consumption. More than half of households raise livestock, with the vast majority keeping pigs, and almost one-third keep goats and chickens (although there are few ruminant animals and no intensive livestock enterprises) (CI Ministry of Agriculture 2001).

However, there is little in the way of agriculture outside of the subsistence sector. Constraints on production of larger volumes of agri-food commodities are a mix of political, economic, geographical and social factors. In the northern group of islands there is a lack of fertile soil, so residents must rely on a diet of fish, coconuts and imported foods. In recent years, pearling and seaweed farming have become economically more important activities than subsistence agriculture (CI Ministry of Agriculture 2001). In the southern group, including Rarotonga, land-use pressures and relatively high production costs have constrained agricultural output. The establishment of the Rarotonga international airport hastened the decline of agriculture on the main island because, first, it consumed approximately 20 per cent of Rarotonga's land mass (Jamieson 2002: 57–8) and, second, it provided an exit point for young males who would otherwise have become local agricultural labourers. In this way, the airport facilitated both migration and remittances. The fortunes of agriculture in the Cook Islands are also inhibited by: a land tenure system in which many small plots of agricultural land are held under multiple 'ownership' (Syed and Mataoi 1993: 103); the conversion of arable land in Rarotonga into land for housing and resort developments; rising prices of seeds and stock feed; food imports tied to development assistance; and difficulties in accessing secure and sustainable export markets.

As a result, agricultural exports from the Cook Islands are minuscule. The few products exported include small amounts of taro, fresh fish and pawpaw, and *maire*, which is used in the making of flower displays and for ceremonial purposes. Indeed, since the 1980s the value of agricultural exports from

Table 6.1 Cook Islands imports of meat and edible offal, 2002

Meats	Quantity (tonnes)	Value (millions)
Beef	114	1.39
Pork	12	0.18
Mutton	165	0.75
Chicken	1,545	1.80
Goat	4	0.06
Total	1,840	4.18

Source: CI Ministry of Agriculture unpublished figures for 2002.

Note: Import values are inclusive of insurance and freight, and are in New Zealand dollars.

the Cook Islands has fallen by 80 per cent. Quarantine restrictions on fresh fruit exports, imposed recently by New Zealand and Australia because of the alleged risk of fruit fly infestation, account for some of this decline. However, the broader picture this reveals is a large and widening trade deficit for food and agriculture. Figures for 2001 show that after machinery and transport, food constituted the country's second largest category of imports (20 per cent of all imports) (CI Statistics Office 2002). In 1990, it was estimated that an average household spent nearly 70 per cent of its food expenditure on imported food (Syed and Mataoi 1993: 101). From a food import bill of NZ$23 million, about one-fifth was spent on meat and edible offals (CI Ministry of Agriculture unpublished figures) and, of this amount, chicken meat represented the largest component by far (Table 6.1).

The Cook Islands' chicken meat commodity complex and changing culinary cultures

The significant importation of chicken meat into the Cook Islands takes place despite plentiful supplies of local birds, notably including wild chickens that roam colourfully around the islands. The wild chicken population tends to provide an opportunistic and supplementary food source for Cook Islanders. According to local informants, large numbers of chickens (and fruit bats) were killed for food in the mid-1990s, following a wave of public sector retrenchments.[1] Since that time, however, the wild chicken numbers have been replenished and, according to many, they now represent a 'pest'. It seems that only a few older people and those without an income continue to utilize this abundant source of free protein. And yet ironically, when Cook Islanders are asked about these birds, they emphasize their superior taste. Local chicken meat is appreciated for its texture, hard bones and strong flavour: it is 'more hearty'. But while island chicken and eggs 'taste beautiful' to many, few have the time to kill and pluck these chickens or to cook them to the point that they are not 'too chewy'. In contrast, imported chicken is quicker to prepare and contains more meat on the bones, a feature that appeals to a people who are proud to be big meat eaters. In addition to these birds, many Cook Islander households keep chicken broods fed from table scraps and supplementary feed.

At the present time there are no commercial poultry operations in the Cook Islands. Economic factors would seem to impose prohibitive barriers against local production. In addition to the high costs of feed and the diseconomies of scale, any local operations would also have to rely on the air-freighted supply of day-old chicks from New Zealand. Nevertheless, the Cook Islands Department of Agriculture is exploring the viability of commercial broiler production in the outer islands, where land is available and the freight costs of imported chicken meat are high (Tamarua 2003).

In the absence of local producers, the provisioning of chicken meat in the Cook Islands depends on imports from overseas producers, which are coordinated by local wholesalers and retailers. Imports are sourced from the New Zealand firm Tegel (owned by H.J. Heinz Co.), the US firm Tyson, the two largest Australian firms Ingham and Bartter, and two smaller long-established Australian family-owned firms Golden Cockerel and Cordina. Tyson supplies large slabs of frozen pieces destined for the restaurant and hotel market, whereas the others supply frozen whole chickens and cartons of chicken pieces to wholesalers and retailers.[2] In addition to several retail stores (CITC, Meat Co and Foodland), chicken meat is also sold through Country Fried Chicken, a New Zealand-based fried chicken chain that has two outlets in Rarotonga. In overall terms, this import trade is coordinated by a small number of Cook Island businessmen with interests in local shipping, airlines and retail points. According to the Cook Islands Deputy Prime Minister: 'Wherever you turn, wherever you look, the same businessmen in Rarotonga are into everything – retail trade, airlines, tourism accommodation, travel agents, restaurants and lately, shipping' (quoted in Mason Tekura'i'moana 2003: 191). Hence, the Cook Islands' chicken meat import complex is enacted and sustained through an interlocking network of overseas corporations and local agents, the latter of which are located within the gravity of political and commercial influence in the island nation.

Evidently, the Cook Islands food import complex is attached to changing culinary cultures. For over half a century, concern has been expressed about the shift to imported food in the Cook Islands. In 1949, an early study:

> documented that although staple foods such as kumara, taro, breadfruit, arrowroot and green bananas were predominant, there was increasing dependence on imported tinned meat, flour, sugar and biscuits. Milk was rarely consumed and if it was, it was reserved for babies or sick

people. Butter, cheese and eggs were rarely eaten. Green vegetables were scarce except for taro and kumara leaves, but only the former (called rukau) was eaten much, and even that was not a daily item. Coconuts were eaten in some form daily. The main source of protein was fish and other seafood. Local fruits were eaten freely when in season. Pawpaw was only eaten by babies and pigs. Breakfast consisted of bread without butter and tea without milk, but it was not uncommon for some to have no breakfast. The main meal was eaten at midday, with the evening meal being similar to breakfast. Fresh meat was a luxury and if eaten was mainly pork.

(Ta'irea 2003: 163–4)

Twenty-five years ago, Fitzgerald (1980: 72) reported that the transition from 'traditional island diets' towards 'New Zealand diets' was characterized by less snacking (as people's time budgets were re-ordered to accommodate Westernized concepts of work and leisure), more regular and larger meals at fixed times, and a greater variety of foods (especially processed foods) acquired through market exchange. Thus, changing culinary cultures are as much about changing practices as they are about the inclusion and exclusion of particular food items. The remainder of the chapter describes the forces that lie behind these transformations, as they apply specifically to chicken meat.

Tourism's impact: chicken meat as 'the authentic island dish'

The South Pacific accounts for only 0.15 per cent of global international tourism arrivals (Hall 1996: 1) but it contributes significantly to local economies in the region. It is a major source of employment and represents an important, although uneven, development option for Pacific Island microstates faced with the economic constraints of size and isolation (Milne 1990; Milne 1992; Hall 1994; Jamieson 2002). In the political context of heightened nationalism since the late 1980s, tourism-led economic development has been a prime focus for the Cook Islands government and private sector. In 2001 the Cook Islands received over 82,000 international arrivals, of which more than 60,000 were vacationers (the remainder mainly comprised return visits by expatriates). New Zealand was the largest single source of arrivals (24,325), followed by Europe (22,817), Australia (11,826) and the US (7,145) (CI Statistics Office 2002: Table 6.1).

Food exercises a major attraction, activity and ritual for many tourists to the Cook Islands. The interface between tourists and local food cultures takes the form of a practice known as 'Island Nights', which is a buffet dinner and a floor show of dance and music. A typical spread of food consists of taro, arrowroot, rukau, clams, octopus, raw fish, prawns, a whole baked fish, numerous salads in mayonnaise, chop suey, coconut with seafood favouring, roasted chicken pieces, a pork roast carved by a chef, and a huge selection of desserts including *poke*, a much-loved dish made with either pawpaw or banana and coconut milk. In addition to the tourists, local people attend Island Nights to celebrate birthdays and for their own entertainment.

The incorporation of imported foods into local cuisine, and their promotion to tourists as an 'island' dish, challenges and reconstructs the imagined notion of 'authenticity' in Cook Islands culinary culture. There is demand by tourists for what they imagine to be a 'local food experience', and the large hotels, resorts and local tours acquiesce by preparing and presenting an array of island-style dishes at tourist feasts. Clearly, a veneer of 'authenticity' masks preparation methods and product sourcing arrangements that bear little resemblance to historically dominant Cook Island fare. Yet what is interesting is that, over time, the reconstructed culinary cultures presented in Island Nights have become accepted and indeed esteemed by local people. Increasingly, trays of roasted chicken pieces are replacing whole island birds cooked in earth ovens (*umu*) at feasts for local people.[3] Instead of three or four chickens cooked underground, a carton of chicken pieces cooked in the electric oven has become acceptable and is even preferred in some instances, because it appears more generous. As one Cook Islander informant commented: '[C]an you imagine if I take only three chickens [to a feast]. They'd say "hey, that's not enough. You should come with one carton" . . . now the expectations have gone too far.'

The commercial incorporation of imported foods into events that have significance for local people is a contemporary manifestation of 'neo-traditional development': the harnessing of custom for commerce (Sahlins 1994: 415). Equally, commercial enterprise is being used to sustain traditional customs; feasts are accompanied by music and dance. This observation accords with Wilk's (2002) analysis of culinary cultures in Belize, another small dependent state. Not only has the performative aspect of the culinary culture been heightened, but national dishes have been reconstructed to appeal to a much wider audience, including tourists, expatriates living in the islands and Cook Islanders living abroad and returning for holidays.

The impact of Cook Islander migration on the culinary culture

In a chapter examining the dynamics of food exchange, Alexeyeff (2004) describes how food provides a powerful material and affective bond among Cook Islanders who are scattered across the Pacific. She tells about the raw produce and cooked meals that those living in the Islands take on their visits to relatives in New Zealand and Australia. She describes the gifts, including food, that accompany them on the return journey as well as the parcels that arrive some weeks later from friends, relatives and even friends of friends. This familial trade in food provides Alexeyeff with evidence to question simplistic understandings of societies reliant on remittances. Because of the two-way flows and the deep emotions attached to the exchanges, she makes a case for the continuation of 'the ancient practice' of reciprocity. Through reciprocal food exchanges, Cook Islander identity and a defined culinary culture are reproduced over long distances.

Our observations provide a coterminous reading of an evolving culinary culture, namely that migration and travel between the Cook Islands and other countries (mainly New Zealand) contribute strongly to lifestyle changes in general and food preferences in particular. Return migrants to the Cook Islands bring with them an array of experiences and perceptions that influence local food systems. In a study of return migrants, but in a very different setting, Teti (1995) describes how Calabrian peasants who migrated to America in the early twentieth century to escape poverty and hunger learnt to eat a variety and amount of food unimaginable in Italy. On their return, the emigrants would 'introduce new foods, break with a diet that had been predominantly vegetarian, overturn the image of the peasant as a sober and frugal man, and show off new behaviour revealing the attainment of a new status' (Teti 1995: 23). Since the 1950s and more widespread affluence in Calabria, Teti observes that there has been a 'a progressive and significant rejection of the "vegetarian regime" and of the Mediterranean diet' reflecting 'the ambiguous relationship to tradition, which at times is asserted, at times denied, at times invented' (Teti 1995: 24).

Similar observations can be made with regard to Cook Island return migrants and travellers. There is an apparent ambiguity to tradition, underlined by the incorporation of fast food and convenience into the Cook Islands' culinary cultures. Specifically, a preference for fast food by youths is an important by-product of commodified travel. This is exemplified by the popularity of KFC, brought back in cooler boxes (what New Zealanders call 'chilly bins') as passenger luggage on flights from New Zealand (Jamieson 2002: 27). Personal importation of such products has important affective meaning (Alexeyeff 2004). According to one Cook Islander informant:

> [it is] part of our culture to bring a contribution to people we are staying with. It is food, but also soap, cleaning things, toilet paper – all the things that contribute to having someone live in your house. The types of food that people take back are meat (steak, chops, chicken) and butter – the things that are hard to get in Rarotonga. We give this rather than money. It is a gesture.

Moreover, the appeal of fast food is attached to Cook Islanders' cultural understandings of food. According to one Cook Islander youth living in Hawaii:

> In the Cook Islands, we associate restaurants with tourists or with locals who hold positions of power and influence. They understand the cuisine

and how to order, table manners, etiquette, how to ask for the bill and pay. Most are happy eating our own foods, with fingers, at home . . . When eating out in the USA, we prefer the simplicity of fast foods or buffet meals with display pictures of the dishes offered.

(Tongia 2003: 320)

Fast food is convenient. As a social construct, convenience is shaping a range of culinary practices of Cook Islanders, whether at home or abroad. In the Cook Islands, convenience is reflected by two recent developments. First, there is an increasing propensity for convenience stores to be co-located with petrol stations. A rapid increase in car and motor cycle ownership (CI Statistics 2002: Table 9.3) is facilitating the emergence of what could be called 'vehicle-centred diets' (Hinde and Dixon 2005). These diets are based on foods prepared in commercial kitchens and consumed on the move, following trends taking hold in developed countries. Second, Rarotonga has seen the arrival of a chain called '6–11', which is open from 6am to 11pm from Monday through Saturday and 7.30am to 9.30pm on Sundays. These stores sell canned foods, bread, milk, fruit and vegetables, soft drinks and ready-to-eat plates of cooked food such as taro, chop suey, 'mayonnaise' (local potato salad) and fried chicken. Increasingly, plates of prepared food purchased direct from these stores are replacing Saturday's half-day preparation of the post-church Sunday meal. A number of Cook Islander informants in Rarotonga expressed the view that they 'can't be bothered cooking' and like the convenience afforded by prepared food being sold by stores with extensive opening hours. On the one hand, investments by fast-food outlets and convenience stores have cheapened the price of prepared foods in the Cook Islands, making them increasingly available and attractive to local residents (CI Statistics Office 2002: Table 3.3). At the same time, the spread of fast food is not only a significant cultural import, but as Ferguson and Zukin (1995) have observed more generally, it has significant political-economic implications in terms of the networks of corporations and individuals that are shaping individuals' food choices in small nations such as the Cook Islands.

In short, the flow of cultural practices based around constructions of 'convenience' is exceedingly important in understanding current transformations in the culinary practices of Cook Islanders, and specifically, the role of chicken meat within these contexts. They are reflective of an ambiguity in relation to tradition, in which practices of generosity, feasting and the special status of chicken within a feast coalesce with recently acquired values of convenience and large meal portions. Rather than Coca-colonization, this signifies a process of 'creolization'. According to Howes (1996: 5), creolization is a paradigm whereby introduced foods are domesticated to fit within different cultural contexts. This is an important distinction, because it acknowledges the creativity and agency of consumers. In discussing the Belizean diet, Wilk

(2002) notes that creolization is achieved through mixing ingredients and cooking styles in new ways, and by substituting imported for indigenous ingredients. Understanding the culinary culture of the Cook Islands in terms of creolization is in keeping with the notion of postcolonial global reality as a history of multiple migrations (Narayan 1997: 187), which is itself long a feature of Cook Islands history.

Conclusion

According to James (1996: 79):

> Trade, travel, transport and technology have all played their part in facilitating a considerable exchange of consumption practices. This brings into question, therefore, the very notion of 'authentic' food traditions, raising doubts as to the validating role food might have with respect to cultural identity.

In the Cook Islands, the influence of tourism and travel emphasizes the set of complex and apparently contradictory processes that surround notions of 'authenticity' and culinary culture. In the day-to-day telling of the story of the Cook Islands to tourists, local people reproduce a sense of what their culture is about. In this way, tourism becomes a vehicle for a celebration of culture, with the Island Night feast playing a central role. Island Nights present a semblance of an authentic feast in both the abundance and generosity of the spread and the style of the food represented, reflecting what Islanders themselves would like to eat. Roast chicken in the feast, irrespective of where the chicken comes from or how it is cooked, continues to play a validating role for cultural identity. Yet in line with international cultural influences on the Cook Islands, aided and abetted by flows of migration and travel from the Cook Islands to New Zealand and further afield, these identities are under continued reformulation. Wages remitted to the homeland facilitates lifestyle changes based on demands for Western-style convenience, albeit creolized as they interact with Cook Islander traditions.

A focus on the cultural economy of chicken meat within these settings provides an entry point from which to observe the contemporary transformations of Cook Islander society. In recent years there have been considerable concerns about nutrition-related health problems in the micro-states of the South Pacific. Populations are becoming obese, with all the attendant health risks of type 2 diabetes, coronary heart disease and skeletal problems (WHO 2003). By replacing traditional sources of nutrition (fish, taro, coconuts, other fruit, baked foods) with modern sources (meat, bread and noodles, fried foods) national populations are at risk of chronic diseases that they will not be able to afford to remedy. Faced with these pressing social problems, analysts and policy-makers often point accusatory fingers at how

Western-style food imports, such as chicken, have replaced local production. Yet as this chapter has also demonstrated, it is as much the importation and creolization of *food practices* as the foods themselves that are contributors to these social outcomes.

Notes

1   This research utilized a series of interviews with Cook Islanders undertaken in the Cook Islands during 2003. Full information on the methods and techniques of these interviews can be obtained by contacting the authors.
2   However, in April 2003, the Cook Islands government suspended imports from the US due to an outbreak of avian influenza in the California flocks.
3   Diminished use of the *umu* (earth oven) appears a result of the social construction of convenience and the changing role of women (Varcoe 1993). While the *umu* is used less for home cooking, it is still used for special events, involving an exclusively island population. The baking of food in the ground and the cooking of goat, the most esteemed meat because of its relative scarcity, distinguishes tourist authentic and indigenous authentic occasions.

References

Alexeyeff, K. (2004) 'Love food: exchange and sustenance in the Cook Islands diaspora', *The Australian Journal of Anthropology*, 15 (1): 68–80.

Appadurai, A. (1990) 'Disjuncture and difference in the global cultural economy', *Public Culture*, 2: 1–24.

Cook Islands (CI) Ministry of Agriculture (2001) *Cook Islands 2000 Census of Agriculture and Fisheries*. Rarotonga: Cook Islands Government.

Cook Islands (CI) Statistics Office (2002) *Annual Statistical Bulletin*. Rarotonga: Cook Islands Government.

Dixon, J. (2002) *The Changing Chicken. Chooks, Cooks and Culinary Culture*. Sydney: UNSW Press.

Ferguson, P. and Zukin, S. (1995) 'What's cooking?', *Theory and Society*, 24 (2): 193–9.

Fitzgerald, T. (1980) 'Dietary change among Cook Islanders in New Zealand', *Social Science Information*, 19 (4/5): 805–32.

Hall, C.M. (1994) *Tourism in the Pacific Rim: Development, Impacts and Markets*. Melbourne: Longman.

Hall, C.M. (1996) 'Introduction: the context of tourism development in the South Pacific'. In C.M. Hall and S.J. Page (eds) *Tourism in the Pacific: Issues and Cases*. London: International Thomson Business Press, pp. 1–15.

Hinde, S. and Dixon, J. (2005) 'Changing the obesogenic environment: insights from a cultural economy of car-reliance', *Transportation Research Part D*, 10 (1): 31–53.

Howes, D. (1996) 'Commodities and cultural borders'. In D. Howes (ed.) *Cross-cultural Consumption: Global Markets, Local Realities*. London and New York: Routledge, pp. 1–16.

James, A. (1996) 'Cooking the books: global or local identities in contemporary British food cultures?'. In D. Howes (ed.), *Cross-cultural Consumption: Global Markets, Local Realities*. London and New York: Routledge, pp. 77–92.

Jamieson, K. (2002) 'In the isle of the beholder: traversing place, exploring representations and experiences of Cook Islands tourism', unpublished thesis, Australian National University.

Mason Tekura'i'moana, J. (2003) 'The cultural influence of corporate power'. In R. Crocombe and M. Crocombe, *Cook Islands Culture*, Rarotonga: Institute of Pacific Studies, pp. 187–97.

Miller, D. (1997) *Capitalism. An Ethnographic Approach*. Oxford: Berg.

Milne, S. (1990) 'The impact of tourism development in Small Pacific island states', *New Zealand Journal of Geography*, 89: 16–21.

Milne, S. (1992) 'Tourism and development in South Pacific microstates', *Annals of Tourism Research*, 19: 191–212.

Narayan, U. (1997) 'Eating cultures: Incorporation, identity and Indian food'. In U. Narayan (ed.) *Dislocating Cultures: Identities, Traditions and Third-World Feminism*. New York: Routledge, pp. 161–219.

Pelto, G. and Pelto, P. (1983) 'Diet and delocalization: dietary changes since 1750'. In R. Rotberg and T. Rabb (eds) *Hunger and History: The Impact of Changing Food Production and Consumption Patterns on Society*. Cambridge: Cambridge University Press, pp. 309–30.

Probyn, E. (1998) 'Mc-Identities: food and the familial citizen', *Theory, Culture and Society*, 15: 155–73.

Sahlins, M. (1994) 'Cosmologies of capitalism: the trans-Pacific sector of "The World System"'. In N. Dirks, G. Eley and S. Ortner (eds) *Culture/Power/History*. Princeton (NJ): Princeton University Press, pp. 412–55.

Syed, S. and Mataio, N. (1993) *Agriculture in the Cook Islands: New Directions*. Rarotonga and Suva: Institute of Pacific Studies and the Cook Islands Centre of the University of the South Pacific.

Ta'irea, K. (2003) 'Kai: the culture of food'. In R. Crocombe and M. Crocombe, *Cook Islands Culture*. Rarotonga: Institute of Pacific Studies, pp. 163–64.

Tamarua, T. (2003) Personal communication, Rarotonga, 14 April. (Mr Tamarua is an employee of the Cook Islands Department of Agriculture.)

Teti, V. (1995) 'Food and fatness in Calabria'. In I. De Garine and N. Pollock (eds) *Social Aspects of Obesity*. Amsterdam: OPA, pp. 3–29.

Tongia, A. (2003) 'USA and Cook Islands culture'. In R. Crocombe and M. Crocombe, *Cook Islands Culture*, Rarotonga: Institute of Pacific Studies, pp. 315–23.

Varcoe, J.K. (1993) 'From the *umu* to the oven: emancipation for the women of the Cook Islands?', unpublished Masters thesis, University of Canterbury.

Wilk, R. (2002). 'Food and nationalism: the origins of Belizean food'. In W. Belasco and P. Scrantin (eds) *Food Nations*. New York: Routledge, pp. 67–91.

World Bank (2002) *Embarking on a Global Voyage: Trade Liberalization and Complementary Reforms in the Pacific*, Pacific Islands Regional Economic Report No 24417-EAP.

World Health Organization (WHO) (2003) 'The food supply'. In *Diet, Food Supply and Obesity in the Pacific*, background document for FAO/SPC/WHO Consultation on Food Safety and Quality in the Pacific, Fiji 11–15 November 2002. Paper prepared by R.G. Hughes.

7 Interpreting the Australian– Philippines food trade in the context of debates on food security*

Stewart Lockie

Introduction

Since 1992, the Australian government has implemented a number of strategies designed to capitalize on the ostensibly natural market provided to Australian food exporters by the rapidly growing populations and economies of Asia (Pritchard 1999). On the surface, increased exports to the region appeared to offer a win-win solution to export-oriented Australian farmers faced with the need to secure new markets in order to arrest declining terms of trade, and to Asian governments and consumers faced with the prospect of procuring enough food of sufficient variety to satisfy both the basic needs of growing populations and the changing tastes of the affluent middle classes. Despite the seemingly straightforward logic of positioning Australia as a 'Supermarket to Asia', in 2002 Philippine farmers hurled rotten vegetables at metropolitan supermarkets in protest at the importation of Australian vegetables (Lacuarta 2002a). Local government representatives claimed that between April and October 2002, produce from the mountainous Cordillera region worth P21 million (approximately US$400,000) was displaced from Metro Manila and Cebu markets by imported vegetables (*Philippine Daily Inquirer* 2002), some 93 per cent of which were sourced in Australia and sold under the misleading local name 'Baguio vegetables' (Lacuarta 2002b).

While the actual impact of legally imported vegetables on Cordillera producers is clouded by the alleged widespread smuggling of vegetables from China and elsewhere (Lacuarta 2002a), it does seem clear there is no direct, or necessarily positive, relationship between the importation of food and the availability, affordability, adequacy or acceptability of food for those most vulnerable to food insecurity. Although Australia and the Philippines are relatively minor trading partners, the food security impacts of this trade are potentially quite significant in a number of ways. First, as an example of

* The author would like to acknowledge the support of the Institute of Philippine Culture at Ateneo de Manila University, and of its Director, Dr Filomeno V. Aguilar, Jr, and coordinator of the Visiting Research Associate Program, Cecilia Honrado. Special thanks must also go to Dr Jeanne Illo for her invaluable guidance and encouragement.

trade that reverses the more widely seen pattern of 'Southern production for Northern consumption', we might reasonably expect the implications for livelihoods and food security to be quite different. Second, as the example of vegetable trade illustrates, the actual volume of trade in any particular commodity, or group of commodities, is less important than the ability of domestic producers in that sector either to compete profitably or shift production to alternative crops. At the same time that the national impact of import competition may seem relatively low, or even positive, the distribution of negative impacts may be concentrated among smaller and more vulnerable groups. Further, as the current dispute between Australia and the Philippines over quarantine requirements and the export of Philippine bananas and other tropical fruits to Australia illustrates (see Department of Foreign Affairs and Trade (DFAT) 2003; Fagan this volume), relationships between domestic agricultural production, trade, politics and food security can only be understood within the full context of conflict over processes of agricultural modernization and trade liberalization. With these processes represented as solutions to myriad social problems, including food insecurity, it is essential that both are assessed for their potential impacts on those most vulnerable to food insecurity in the Philippines and elsewhere.

Unpicking the entire web of relationships between food security, livelihoods, agricultural modernization, trade liberalization, agrarian politics, Australia–Philippines trade flows and so on is beyond the scope of this chapter. The chapter offers, therefore, an initial exploration of trade relationships between the two countries with a view to drawing out some of the implications of the liberalization and modernization agendas for food security and rural livelihoods. It concludes by suggesting a research agenda that might allow us to draw firmer conclusions about these implications.

Food availability and adequacy in the Philippines

Although there is probably sufficient food within the Philippines to meet basic needs, the typical Filipino diet is grossly inadequate in energy and nutrients (Briones *et al.* 1999; Bayanai and Marchesich 2001). In 1998, 31.8 per cent of pre-school children were underweight for age, 32.0 per cent were stunted (under-height for age), 6.6 per cent were wasted (underweight for height) and 1.0 per cent were overweight (Bayanai and Marchesich 2001). At the same time, 19.8 per cent of adolescents and 13.2 per cent of adults were underweight and energy deficient. Women – especially those who were pregnant or lactating – were found to be particularly vulnerable. Iron-deficiency anaemia affected 30.6 per cent of the population while significant numbers were affected by vitamin A and iodine deficiencies. A major cause of malnutrition in the Philippines is poverty, with some 37.5 per cent of the population unable to meet their most basic food and other needs in 1997 (Bayanai and Marchesich 2001). Using a different methodology (based on expenditure rather than income), the Philippine Human Development

Report (Human Development Network and United Nations Development Programme 2002) reports that this situation deteriorated further between 1997 and 2000 with the number of impoverished Filipinos increasing from 25.1 per cent to 27.5 per cent of the total population. The 1991 Family Income and Expenditure Survey showed that urban families spent up to 64.6 per cent of their income on cereals and rural families up to 66.6 per cent (Mariano 1996).

Poverty is closely related to reliance on the agricultural sector with 65.6 per cent of the poor population residing in rural areas in 1994 increasing to 71.5 per cent in 1997 (Bayanai and Marchesich 2001). In a survey of farm workers in the sugar industry in 1999, 90 per cent believed that food consumption in their households had declined since 1995 due to high prices, low wages and underemployment (Tujun 2000). The most food-insecure households nationally include upland farmers, lowland crop farmers, agricultural workers, subsistence fishermen and the urban poor (Briones *et al.* 1999).

Factors likely to place continued pressure on the ability of the Philippines to meet basic food needs include:

* rapid population growth – the Philippine population increased from 39 to 77 million between 1972 and 1991 (FAO 2003). It is projected to reach 115 million in 2025 (Hossain and Sombilla 1999) and thence to continue growing well into the twenty-first century (Paunlagui 1999);
* escalating food needs – national self-sufficiency in rice production in the year 2010 would require an increase in production of nearly 50 per cent over 1990s levels (Hossain and Sombilla 1999);
* limited scope to expand production – nearly all available arable land on the Philippine archipelago of only 300,000 square kilometres is already in agricultural use (FAO 2003), with 90 per cent of land suitable for cultivation of high-yielding rice varieties already used for this purpose (Estudillo *et al.* 1999; Hossain and Sombilla 1999).

Competing visions of food security

Despite its status as a chronically food-insecure country, the meaning and implications of food security in the Philippines are fiercely contested (Bello 1997). Among the multitude of positions on food security two broad schools of thought may be discerned. The first – which is currently ascendant in national and international policy – promotes a minimalist view of food security as the availability and affordability of nutritionally adequate and culturally acceptable food (Cabanilla 1999). According to the minimalist view, the origin of food is immaterial so long as it meets the needs of consumers. Not surprisingly, this view is promoted by those governments and agencies also responsible for championing trade liberalization and the modernization of traditional agricultural sectors (Madeley 2000).

Liberalization and modernization are proposed by such agencies as the anti-dotes to chronic food insecurity by shifting production of staple foods to those countries and regions in which resources can be utilized most effi-ciently, lowering the price of food for consumers, and boosting incomes in those agricultural regions in which resources may be more effectively used to grow higher-value alternative crops (Bello 1997; Madeley 2000). The notion of Australia as a 'Supermarket to Asia' fits very comfortably with this vision for food security since, even in the event that markets for those products sold by Australia are oversupplied, any comparative advantage held by Australian producers will merely provide market signals to Philippine producers that they should redeploy their resources elsewhere. Trade deficits, however, remain problematic since they undermine long-term capacity to afford food imports.

The second school of thought on food security is centred on the concept of 'food sovereignty'. While this has become a marginalized position under the tide of neo-liberal policies – and is regarded as simply incorrect by many economists (e.g. Cabanilla 1999) – it is vigorously promoted by non-government organizations and farmer groups. Within this school there are a number of emphases that reflect the diverse coalitions of opponents to wholesale trade liberalization. For some, food security is tied intrinsically to self-sufficiency and the capacity of Philippine agriculture to meet domestic demand for all staple foods, thus buffering domestic producers and consumers from world market volatility (IBON 1999a; Rosario-Malonzo 2001). Estudillo *et al.* (1999), for example, argue that as a predominantly subsist-ence crop (less than 5 per cent of global production is traded internationally) the world supply of rice is highly unpredictable and dependence on it unduly risky. For others, the core issues centre more on who controls the food supply and the livelihoods of those dependent on it (Arao 2000a). While trade may not necessarily be inconsistent with this conceptualization of food sover-eignty, the specific approaches that have been taken to trade liberalization and agricultural modernization in the Philippines are seen to have trans-ferred control of Philippine agriculture to transnational corporations (TNCs) and agencies (such as the WTO) while undermining the livelihoods of the majority of Filipino farmers and doing little to lower the cost of food for consumers. Moves to further embed the influence of TNCs through contract-growing arrangements, the introduction of plant variety rights legislation, promotion of high-input Green Revolution technologies and so on are all seen as highly problematic. Overall, the food sovereignty approach to food security does not preclude a role for Australian imports – either of food or production technologies – but limits these to sectors and technologies that do not threaten the ability of Filipinos to decide how they will meet their own basic needs.

Irrespective of the position taken within these debates, the four elements of food security offered by the minimalist position – availability, afford-ability, nutritional adequacy and cultural acceptability – offer essential criteria by which to evaluate the impact of both the liberalization and sovereignty agendas. Before examining the possible impacts of Australian trade in particular, this chapter will consider Philippine performance against these criteria during the period of modernization and trade liberalization following the Green Revolution of the 1970s.

The Green Revolution and trade liberalization

Moves to modernize and liberalize Philippine agriculture have been undertaken more or less simultaneously since the 1960s. The International Monetary Fund (IMF) first imposed trade reform in 1962 and again in 1973 (Scipes 1999; Guzman 2000a). In 1974, with World Bank funding, the Philippine government began promoting more vigorously the adoption of Green Revolution technologies – including high-yielding varieties (HYVs) of rice and corn – through the provision of credit, land reform and cooperative programmes (Herdt 1987; Estudillo *et al.* 1999). According to critics, the dependence of HYVs on optimum growing conditions provided by irrigation, synthetic inputs (fertilizer and pesticides) and favourable seasonal conditions resulted in disappointing and erratic results for poor farmers (Lim 1996). This, they argue, contributed to a cycle of loan defaults, increasing indebtedness and falling yields (Lim 1996). By contrast, Herdt (1987) argues that supporters and detractors of Green Revolution technologies alike have oversimplified and exaggerated their impacts – both positive and negative – but that by the mid-1980s across Asia the empirical literature tended to suggest that HYVs had been adopted quite evenly among farmers of all size groups and had led to increased output and modest increases in labour demand (see also Estudillo *et al.* 1999). Nevertheless, a significant proportion of the improved output from Philippine agriculture in the early years of the Green Revolution could be accounted for by an expansion of agricultural land use (Figure 7.1). Other factors included agricultural mechanization and irrigation infrastructure development as well as improvements in seed technology (Estudillo *et al.* 1999).

Moreover, these increases in agricultural production have been sufficient merely to avoid further declines in per capita food production (Figure 7.2). Importantly, these trends have not been uniform across agricultural crops. Rice yields – which are of particular importance due to the status of rice as a staple crop – increased from 1.3 to 2.9 tonnes per hectare between 1965 and 1994 (Hossain and Sombilla 1999). The slowing rate of increase since the mid-1980s can be accounted for by a failure to lift substantially the yield potential of HYVs (Estudillo *et al.* 1999) as well as reduced public expenditure on maintenance and expansion of irrigation, limited availability of land suitable for modern high-yielding varieties, and encroachment into rice-producing areas of industrial and residential land uses (Estudillo *et al.* 1999; Hossain and Sombilla 1999).

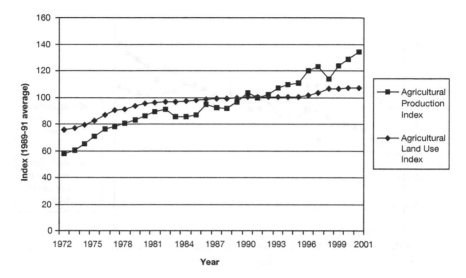

Figure 7.1 Agricultural Production and Land Use Indexes, the Philippines, 1972–2001

Source: FAO 2003.

Notes: Agricultural Production Index: production quantities of each commodity are based on the sum of price-weighted quantities of different agricultural commodities produced after deductions of quantities used as seed and feed weighted in a similar manner. The resulting aggregate represents disposable production for any use except as seed and feed. To obtain the index, the aggregate for a given year is divided by the average aggregate for the base period 1989–91.

Agricultural Land Use Index: for each year the land area devoted to agricultural production is divided by the average land area for the base period 1989–91.

National debt crisis resulted, in 1979, in the imposition of one of the IMF's first Structural Adjustment Programs requiring tariff reductions, import liberalization and indirect tax reform (Ofrenco 1996; Scipes 1999; Guzman 2000a). While there is insufficient space here to detail ongoing programmes of trade liberalization through the 1980s and early 1990s, it is telling that by the time the General Agreement on Tariffs and Trade (GATT) was ratified in 1994 the Philippines had half the allowable rate of agricultural price and production subsidization of 10 per cent of production value (Guzman 2000a).

From the first IMF loan in 1962 onwards, trade reform in the Philippines has resulted in the expansion of export plantation crops such as banana and pineapple (Guzman 2000a). According to critics, often this forced producers of staple crops – including subsistence farmers – into marginal lands (Atienza 1992). Following implementation of the GATT, land conversion for export crops and industry was pursued more deliberately. The Medium-Term Agricultural Development Plan (MTADP 1993–8) focused on the development of export-competitive high-value crops (HVCs) such as asparagus, zucchini, tomato, garlic, onion, cauliflower, carrot, celery, cabbage,

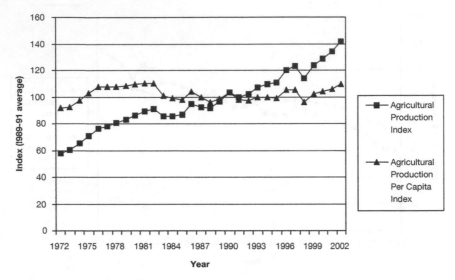

Figure 7.2 Agricultural Production Index in total and per capita terms, the Philippines, 1972–2002

Source: FAO 2003.

Notes: Agricultural Production Index: production quantities of each commodity are based on the sum of price-weighted quantities of different agricultural commodities produced after deductions of quantities used as seed and feed weighted in a similar manner. The resulting aggregate represents disposable production for any use except as seed and feed. To obtain the index, the aggregate for a given year is divided by the average aggregate for the base period 1989–91.

Agricultural Production Per Capita Index: for each year the aggregate quantity of agricultural commodities is divided by the total population of persons and thence divided by the average production per capita for the base period 1989–91.

castor bean, cut flowers and so on, with a goal of reducing the land devoted to food grains from 5 million to 1.9 million hectares (Ofrenco 1996). In addition to providing credit, the Philippine government reconfigured its agrarian reform programme to facilitate contract growing, joint venture and leasing relationships between farmers and corporate agri-businesses (Tujun 2000). Despite this, the goals of the MTADP have not been achieved. Instead, from 1983 to 2002 the area of rice harvested increased from 3.1 million hectares to 4 million hectares (FAO 2003), demonstrating the limited ability of poor peasant farmers to invest in HVCs (Tujun 2000). Further, the only crop for which there was a significant increase in export volumes over the preceding decade was the banana crop, with exports more than doubling to over 2 million tonnes, worth almost US$300 million in 2001 (FAO 2003). Exports of coconut were static at around 2 million metric tonnes despite increasing production, while exports of vegetables were negligible (FAO 2003). This reflected the dominance of the vegetable sector by small rice growers seeking to supplement income by supplying vegetables to the local market (Guzman 2000b) and the collapse of farmgate prices due to competition from imported and smuggled vegetables (Aquino 2003). Sugar, meanwhile, lost its status as an export crop and became subject to net imports which peaked at over half a million tonnes in 1996 (FAO 2003). The staple crops rice and corn both, controversially, registered significant increases in imports (FAO 2003).

From the perspective of food security, it is also important to note that direct government intervention in Philippine agriculture prior to the GATT occurred primarily through the procurement (both domestically and internationally), warehousing and distribution of basic food items to prevent price manipulation by merchants (Tujan 2000). But, by 1989, government procurement was reduced to 2.2 per cent of the domestic rice crop while imports were increased (Guzman 2000a). Also, by this time, rice production in the Philippines had lost its comparative advantage with imported rice (Estudillo *et al.* 1999) due to the relative inefficiency and cost structure of Philippine producers (Arao 2000b; Madeley 2000). Yet despite the availability of cheaper imported rice, domestic retail rice prices increased during the 1990s (Estudillo *et al.* 1999). This resulted, according to Estudillo *et al.* (1999), in higher profit margins for Filipino rice growers. However, it is important to note that Estudillo *et al.* (1999) base their conclusions on a sample of farmers drawn from rice-growing areas relatively well endowed with irrigation and transport infrastructure. Other authors assert that the beneficiaries primarily were private traders and cartels that took control over distribution, processing and retailing and inflated retail prices (Arao 2000b; Tujun 2000).

To summarize, attempts to modernize and liberalize Philippine agriculture have seen retail food prices rise at the same time that productivity gains within agriculture have remained insufficient to maintain comparative advantage. The point here is not to suggest that trade liberalization is solely responsible for the ills of Philippine agriculture but that it appears, by itself, inadequate to address them.

Impacts of Australian trade on Philippine producers and food availability

Australia was the Philippines' fourteenth largest export destination in 2002 (accounting for 1 per cent of total merchandise exports), and its thirteenth largest source of imports (accounting for 1.6 per cent of total merchandise imports). As Figure 7.3 shows, Australia has traditionally enjoyed a substantial trade surplus with the Philippines, much of which can be accounted for by agricultural products including milk, beef and live cattle. The dramatic reduction in this surplus that is evident from 2001 onwards is due in no small way to the export success of the Philippine banana. For despite the fact that not a single banana has been traded between Australia and the Philippines, their influence in trade politics between the two countries has been immense. Australia's refusal to grant import licences for Philippine

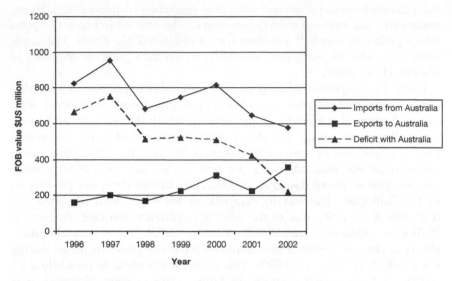

Figure 7.3 Philippines–Australia merchandise trade, 1996–2002

Source: Department of Trade and Industry, Philippines, 2003.

Note: FOB = 'free-on-board'.

bananas (as well as pineapples and mangoes) due to quarantine concerns has contributed to a situation where the Philippines government and national trade organizations have publicly expressed the view that they may shift sources of import supply away from Australia, especially dairy (Table 7.1) (see also Fagan, this volume).

The main concern of this chapter is, of course, not whether this balance of trade is 'fair', but the effect it is likely to have on food security. As argued above, the critical questions revolve, therefore, around the impact of this trade on the availability, affordability, nutritional adequacy and cultural acceptability of food within the Philippines. Further, given the particularly rural profile of poverty in the Philippines, it is especially important to consider the impact of Philippines–Australia trade on rural incomes and livelihoods. Leaving aside, for a moment, the issue of scale (and the obvious contention that as a relatively minor trading partner it is the Philippines trade with other countries that will have the greatest impact on the lives of the poor) it is possible to make three key observations regarding impacts associated with the improvement of Philippine export performance to Australia, the refusal of Australia to accept imports of Philippine fruit and the ability of Filipino producers to compete with Australian producers in markets for HVCs:

1 *The linkage between the balance of trade and food security.* Improvement in Philippine export performance to Australia in recent years may be seen –

Table 7.1 Australian food exports to Philippines, 1996 and 2002 (US$ million)

	1996	2002	Change 1996–2002
Processed dairy	189.8	147.7	−42.1
Processed cereals	27.0	25.7	−1.3
Unprocessed cereals	15.2	10.8	−4.4
Fresh meat	29.8	22.2	−7.6
Processed sugar	18.5	2.1	−16.4
Unprocessed sugar	13.9	4.1	−9.8
Live animals	0.0	13.5	13.5
Animal feeds	5.4	9.2	3.8
Processed vegetables	3.3	7.3	4.0
Fresh vegetables	0.4	0.6	0.2
Cocoa	1.4	5.9	4.5
Beverages	1.9	2.2	0.3
Processed fruit	1.6	1.3	−0.3
Fresh fruit	1.8	0.5	−1.3
Confectionery and honey	0.5	2.4	1.9
Other	7.4	10.9	3.5
Total	318.0	266.4	−51.6

Source: Department of Trade and Industry (Government of the Philippines) 2003.

from the perspective of the minimalist vision of food security – to have increased food security at a national level by reducing the trade deficit with Australia and improving national capacity to afford food imports. However, dominated as it is by industrial manufactures, this expansion of exports is unlikely to have offered improved livelihood opportunities to rural Filipinos counted among those most vulnerable to food insecurity and, if anything, it may have contributed to continued rural–urban migration.

2 *The linkage between high-value crops and food security.* The restrictions placed by Australia on the importation of Philippine bananas, mangoes and pineapples has been presented by the Philippine government as a barrier to regional development, poverty alleviation and political stability on the island of Mindanao where the majority of export bananas and other fruits are grown (Chong 2002). According to this argument, by providing villagers with employment and incomes the export fruit industry reduces the likelihood they will join with Muslim rebels in their struggle against government forces (see Fagan this volume). Unfortunately, there is limited evidence that involvement in the export fruit industry improves the economic situation of anybody other than a limited number of local elites and transnational firms. Davao Province (the centre of export banana growing and trade on Mindanao) is predominantly Roman Catholic. Provinces comprising the Muslim Mindanao Autonomous Region have negligible involvement in the export fruit growing businesses, instead predominantly growing rice, corn and copra. They are also among the ten poorest in the Philippines (Human

Development Network and United Nations Development Programme 2002). Although poverty in Davao Province is dramatically lower than in the Muslim Mindanao Autonomous Region, it still increased slightly between 1997 and 2000 from 26.2 to 27.3 per cent (Human Development Network and United Nations Development Programme 2002). This occurred at the same time as banana production and exports were accelerating some 30 and 40 per cent respectively, with Davao accounting for roughly a third of total national exports. While there is a need for care in imputing direct lines of causality between these data, we would expect to find – were the relationships between banana exports, regional development, employment opportunities and so on as direct as the Philippine government has claimed – some evidence of poverty reduction in Davao. Instead, the overall level of poverty in Davao has deteriorated slightly and remained more or less identical to the national average. At the same time, plantation fruit-growing industries remain the focus of considerable criticism over manipulation of the Comprehensive Agrarian Reform Program and exploitative employment and contract farming arrangements (Atienza 1992; Batara 1996; IBON 1998; Feranil 1999; Homeres *et al.* 2000).

3 *The linkage between food imports and food security.* Australian imports compete directly with a number of Philippine agricultural sectors that – while oriented to the production of cash crops for domestic and international consumption – are ill equipped to deal with such competition (Madeley 2000). With evidence that cheaper wholesale prices due to import competition do not lead necessarily to cheaper retail prices, the real issue here is what they do to farm profitability. In the case of vegetables, official statistics provide a misleading picture of the volume of imports competing with local produce due to the large volume of vegetables that are either smuggled into the country or incorrectly declared at customs (Aquino 2003). While it is impossible to quantify the exact impact of Australian imports on domestic prices, it is telling that the Philippine vegetable industry – despite its promotion by government as an HVC – is understood by government officials and farmer groups alike to be in a state of deep crisis due to its incapacity to compete with cheaper and higher quality imports (IBON 1999b; Aquino 2003; Escandor and Pelayo 2003). Contrary to some claims that imports actually improve food security by supplying markets such as the food service sector (hotels and restaurants) that demand vegetable varieties and quality standards which cannot locally be met – thus freeing Philippine growers to supply domestic food needs – such markets appear freely to switch between local and imported product on the basis of price as well as quality (Aquino 2003). The critical issue here is not whether Philippine producers supply hotels or villages but, again, whether they receive sufficient return on their investment in a crop to meet the livelihood needs of themselves, their families and their workers. Clearly, this is not the case, either for vegetables or for a variety of other crops including sugar, rice and corn (Aquino 1998; IBON 1999c).

Farmers do not have the luxury of opting to supply a local market untouched by the influence of global trade. Instead, Philippine peasant farmers face what may be described as a triple whammy. As land is converted to HVCs it is reconcentrated, and corporate land schemes are put in place that lock farmers into contract growing and credit schemes. Those farmers who move into HVCs are trapped between rising production costs, lower prices, dependence on the infrastructure and technical assistance provided by agri-business, and indebtedness. Meanwhile, cheaper imports of staple crops undermine those remaining in traditional crops such as rice (Guzman 2000a). Contrary to the proposition of neo-liberal political orthodoxy that increasing exposure to international competition will encourage Filipino producers to shift into those enterprises in which they have a comparative advantage – thus securing long-term productivity and profitability – excessive costs imposed by poor post-harvest and transport infrastructure, high input prices, extreme interest rates, poor land tenure, contractual obligations to transnational agri-business firms and internal corruption place major constraints on their capacity to do so (Briones *et al.* 1999; Cabanilla 1999; Costales 1999; Estudillo *et al.* 1999; IBON 1999b).

Conclusion

The secondary data presented in this chapter do not yet tell us the full story of Australia–Philippine trade and its effects on the livelihoods and food security of Filipino farmers, but they do give us some insight into the exposure of Philippine farmers to the competitive pressures of the global marketplace and point towards constructive research foci. While many would construe the liberalization of food trade as positive – encouraging Philippine producers to abandon commodities that may be produced more efficiently elsewhere – a range of factors outside the control of small peasant farmers limit their ability to find alternative market niches with which to secure reasonable livelihoods for themselves, their families and other rural workers. This suggests that something of a contradiction may be found within the minimalist vision of food security as availability, affordability, adequacy and acceptability. At the same time that this vision is put forward in association with a trade reform and export agenda to discredit notions of food sovereignty or self-sufficiency, experience to date with trade liberalization and the expansion of tropical fruit exports suggests that little has been achieved either to boost rural incomes or to lower food prices.

A food self-sufficiency perspective might suggest that any attempt seriously to address food insecurity must itself address the root causes of declining comparative advantage in staple crop production. Thus, Estudillo *et al.* (1999) argue for greater investment in the development and adoption of new varieties with higher yield potentials together with improvements in irrigation infrastructure and crop management, but food sovereignty proponents point towards the relations of production under which that food is grown, a consideration necessary to ensure that small farmers and rural workers have control over their labour and resources and receive a fair return for them. Not only does the export-led strategy for regional development fail adequately to consider such relations of production, it fails also to consider the opportunity cost of ignoring other potential development paths. The basis on which food sovereignty would be built is proposed by Philippine farmer groups and NGOs as genuine land reform, infrastructure investment, affordable credit, farmer-led research, protection of farmers' and public intellectual property, and so on (*Farm News and Views* 1996). All of these are complex issues deserving considerable scholarly attention. The expansive construction of cross-continental food chain studies presented in this book offers a methodological framework through which to allocate such attention in a manner that draws these issues together rather than dealing with them independently. Critically, such an approach would allow us to look more closely at the distribution and concentration of impacts, positive and negative, that accompany international food trade.

References

Aquino, C. (1998) 'Changing the rules of the game: the 1999 review of the GATT-UR Agreement on Agriculture and the Future of Filipino Farmers', *Philippine Peasant Institute Briefing Paper* 6 (4): 1–27.

Aquino, C. (2003) 'The vegetable industry: almost comatose', *Farm News and Views*, 1st Quarter 2003: 1–12.

Arao, D. (2000a) 'Globalization and food security', *IBON Special Release*, 49.

Arao, D. (2000b) 'Impact of the WTO on the Philippine cereals sector'. In A. Tujun (ed.) *The Impact of the WTO Agreement on Agriculture*. Manila: IBON Books, pp. 93–116.

Atienza, J. (1992) *Del Monte Expansion: Whither the Small Farmers and Agrarian Reform?* Manila: Institute on Church and Social Issues, Ateneo de Manila University: PULSO Monograph 10.

Batara, J. (1996) *The Comprehensive Agrarian Reform Program: More Misery for the Philippine Peasantry*. Manila: IBON Books.

Bayani, E. and Marchesich, R. (2001) *Nutrition Country Profile of the Philippines*. Rome: Food and Agriculture Organization of the United Nations.

Bello, W. (1997) 'Strategic policy for food security', *Public Policy* 1 (1): 90–112.

Briones, R., Corcolon, R., Sumalde, Z. and Villancio, V. (1999) 'Food security: household perspective'. In L. Cabanilla, and M. Paunlagui (eds) *Food Security in the Philippines*. Manila: Institute of Strategic Planning and Policy Studies and University of the Philippines Centre for Integrative and Development Studies, pp. 65–79.

Cabanilla, L. (1999) 'Achieving food security: some critical points to consider'. In L. Cabanilla and M. Paunlagui (eds) *Food Security in the Philippines*. Manila: Institute of Strategic Planning and Policy Studies and University of the Philippines Centre for Integrative and Development Studies, pp. 1–20.

Chong, F. (2002) 'Bending over to please everyone', *The Australian* 29 April: 32.

Costales, A. (1999) 'The state of rural road infrastructure'. In L. Cabanilla and M. Paunlagui (eds) *Food Security in the Philippines*. Manila: Institute of Strategic Planning and Policy Studies and University of the Philippines Centre for Integrative and Development Studies, pp. 159–84.

Department of Foreign Affairs and Trade (DFAT) (Australia) (2003) *Australia and WTO Dispute Settlement: Monthly Bulletin*, August. Online: www.dfat.gov.au (accessed 24 September 2003).

Department of Trade and Industry (Philippines) (2003) 'Summary of merchandise imports and exports by country'. Online: http://tradelinephil.dti.gov.ph (accessed 5 December 2003)

Escandor, J. and Pelayo, A. (2003) 'RP seen losing "vegetable war"', *Philippine Daily Inquirer*, 26 March: B6.

Estudillo, J., Fujimura, M. and Hossain, M. (1999) 'New rice technology and comparative advantage in rice production in the Philippines', *The Journal of Development Studies* 35 (5): 162–84.

Farm News and Views (1996) 'Globalisation: trading away liberty through liberalisation', 9 (3 & 4): 1–2.

Feranil, S. (1999) *The Philippine Banana Industry: Confronting the Challenge of Agrarian Reform*. Quezon City: Philippine Peasant Institute and Philippine Network of Rural Development Institutes.

Food and Agriculture Organization (FAO) (2003) 'FAOSTATS agricultural data'. Online: www.faostats.org (accessed various dates).

Guzman, R.B. (2000a) 'The GATT agreement of agriculture: final blow to Philippine farms?' In A. Tujun (ed.) *The Impact of the WTO Agreement on Agriculture*. Manila: IBON Books: pp. 27–64.

Guzman, R.B. (2000b) 'The impact of the GATT-WTO on the Philippine vegetable sector'. In A. Tujun (ed.) *The Impact of the WTO Agreement on Agriculture*. Manila: IBON Books, pp. 153–94.

Herdt, R. (1987) 'A retrospective view of technological and other changes in Philippine rice farming', *Economic Development and Cultural Change* 35 (2): 329–49.

Homeres, G., Mendoza, M. and Yumol, M. (2000) *The Struggle of Small Banana Growers: Hard Won Gains*. Quezon City: Philippine Peasant Institute.

Hossain, M. and Sombilla, M. (1999) 'World grains market: implications for a food security strategy'. In L. Cabanilla and M. Paunlagui (eds) *Food Security in the Philippines*. Manila: Institute of Strategic Planning and Policy Studies and University of the Philippines Centre for Integrative and Development Studies, pp. 21–48.

Human Development Network and United Nations Development Programme (2002) *Philippine Human Development Report 2002*. New York and Geneva: United Nations.

IBON (1998) *Contract Growing: Intensifying TNC Control in Philippine Agriculture*. Manila: IBON Books.

IBON (1999a) 'The Philippines food program: food for whom?', *IBON Facts and Figures* 22 (7–8): 1–11.

IBON (1999b) 'GATT and vegetable farming: counting the costs', *IBON Facts and Figures* 22 (17–18): 1–15.

IBON (1999c) *Impact of the GATT on the Philippine Sugar Industry*. Manila: IBON Special Release 45, June.

Lacuarta, G. (2002a) 'Imported veggies are cheaper, no thanks to WTO', *Philippine Daily Inquirer*, 2 October: Section 1.

Lacuarta, G. (2002b) '"Baguio vegetables" come from Australia', *Philippine Daily Inquirer*, 30 October: Section 1.

Lim, J. (1996) 'Issues concerning to three major agricultural crops and GATT'. In *The General Agreement on Tariffs and Trade: Philippine Issues and Perspectives*. Quezon City: Philippine Peasant Institute, pp. 29–86.

Madeley, J. (2000) *Hungry for Trade: How the Poor Pay for Free Trade*. London: Zed Books.

Mariano, A. (1996) 'Threatening food self-sufficiency: GATT's impact on the grains industry'. In *The General Agreement on Tariffs and Trade: Philippine Issues and Perspectives*. Quezon City: Philippine Peasant Institute, pp. 87–116.

Ofrenco, R. (1996) 'GATT and the non-traditional exports: global farming for whom?', in *The General Agreement on Tariffs and Trade: Philippine Issues and Perspectives*. Quezon City: Philippine Peasant Institute, pp. 117–139.

Paunlagui, M. (1999) 'Population and food requirements'. In L. Cabanilla and M. Paunlagui (eds) *Food Security in the Philippines*. Manila: Institute of Strategic Planning and Policy Studies and University of the Philippines Centre for Integrative and Development Studies, pp. 49–64.

Philippine Daily Inquirer (2002) 'Importers agree to buy local veggies', 31 October: Section 14.

Pritchard, B. (1999) 'Australia as the supermarket to Asia? Governments, territory and political economy in the Australian agri-food system', *Rural Sociology* 64 (2): 284–301.

Rosario-Malonzo, J. (2001) 'Agreement on Agriculture: endangering food security', *IBON Facts and Figures* 24 (16): 1–11.

Scipes, K. (1999) 'Global economic crisis, neoliberal solutions, and the Philippines', *Monthly Review* 51 (7): 1–14.

Tujun, A. (2000) 'The impact of the WTO on food security in the Philippines: case study: Philippine sugar farm sector'. In A. Tujun (ed.) *The Impact of the WTO Agreement on Agriculture*. Manila: IBON Books, pp. 65–92.

8 Inscribed bodies within commodity chains

Global wheat and local insecurity

Jörg Gertel

In 1921 some 36 firms accounted for 85 per cent of the United States' wheat exports; by the end of the 1970s just six companies – Cargill, Continental Grain, Louis Dreyfus, Bunge, Andrea & Co and Mitsui/Cook – exported 96 per cent of all US wheat, 95 per cent of its corn, 90 per cent of its oats and 80 per cent of the nation's sorghum. The top five companies also handled 90 per cent of the Common Market's trade in wheat and corn, 90 per cent of Canada's barley exports, 80 per cent of Argentina's wheat exports and 90 per cent of Australia's sorghum exports. Together, the aforementioned six companies accounted for over 60 per cent of the world's grain traffic, including shipments under food assistance programmes.

(Krebs 1992: 303)

An old man, his 35-year-old wife and three children of nine, six and three live in the outskirts of Cairo, Egypt. They inhabit a dark room below the street level; the room stretches about two by two metres and offers space for one bed only. There is no fresh air, no electricity, no water supply and no bathroom available. The family does not have enough money to cook meals. They eat bread and whatever else they can afford. They live on what they receive from the mosque and the gifts of other people – 'from the mercy of God' as they put it. Along with the price increases of the staple foods during the years of economic 'liberalization' they had to substitute fruit and vegetables increasingly with bread, but even wheat is getting more expensive. Now, the children are sick more often and can hardly concentrate for a longer period.

(Gertel 2002a: 35)

Introduction

Aysh – synonymous with life and bread in Egyptian Arabic – is no longer determined exclusively at the local level in Cairo. Within the context of globalization the 'social logic of localities' (Watts 1989: 3) is becoming increasingly penetrated and shaped by quite distant forces. This is what

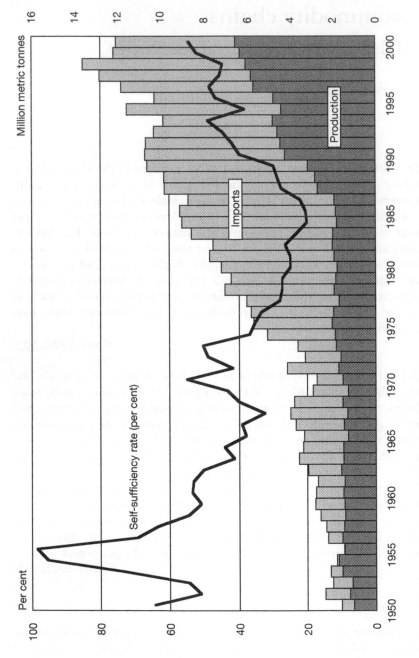

Figure 8.1 Egypt: wheat self-sufficiency rate, 1950–2000

Source: Author.

Giddens describes as the dislocation of space from place (1992: 19). Thus, the most basic needs, such as local food security, are subject to new transformations and risks. This chapter argues that the major driving forces of these changes in Egypt are the large-scale economic liberalization measures aiming to empower market forces while reducing governmental regulations. Pressured by structural adjustment programmes, the public food-provisioning system has become increasingly dismantled since the late 1980s, with particularly dramatic consequences for the poor. In the following, one central aspect of this new food insecurity is examined, namely, the privatization of the global wheat commodity chain and its consequences for local livelihood security.

Food security in Egypt still has a high political priority. This is primarily due to the limited cultivable land – only about 3.5 per cent of the country's area can be used for agriculture. In addition, the Egyptian population, comprising about 70.7 million people (2002), is growing by over 1 million people per year. Therefore, domestic food production is not capable of satisfying the increasing demand, especially for wheat, the staple food in Egypt; the country has thus to cope with a huge food gap (Figure 8.1). Egypt counts as one of the world's most import-dependent countries, and it has received, at times, more food aid than Bangladesh and India (World Bank 1992: 224). But only when Egypt's vulnerable economic situation and the conditions of social inequality are considered does this phenomenon reveal its tremendous problematic significance: between 22.5 per cent (Institute of National Planning (INP) 1996: 25) and 49.0 per cent (Korayem 1996: 21) of the urban population are classified as poor, depending on the respective definition. Moreover, even conservative estimates show that the group of urban poor increased alongside economic liberalization, from 18.2 per cent to 22.5 per cent between 1981 and 1997 (Adams 2000: 263).

In Cairo this precarious situation became dramatically visible with the 'IMF Bread Riots' in 1977. Caused by the policy of the International Monetary Fund, the Egyptian government triggered mass protests with its reduction of food subsidization. The protests ended only when the austerity measures were dropped. In 2004 the situation is apparently relaxed, but only when viewed from the surface. The Egyptian metropolis not only contains an obviously larger population – about 14 million – but it also houses an increasing number of poor people, who are exposed to declines in their entitlement to food.

Conceptualizing risks in cross-continental food chains

In order to investigate the complex relations between the recent restructuring of global wheat chains and local food insecurity in Cairo, this chapter aims to open up a new perspective on commodity chain approaches. Commodity chain and *filière* approaches focus on the flow of specific commodities (Raikes *et al.* 2000). They are primarily concerned with the system of activities and the network of businesses involved in the production, distribution and delivery of products or services to customers (Gereffi and Korzeniewicz 1994). The conditions of social reproduction, and in particular the exposure of consumers to risks – such as market or entitlement failures – are, however, beyond the scope of these approaches.[1]

Starting from here, my argument is based on the following assumption: as economy is embedded within society (Granovetter 1985), the understanding of a specific commodity (such as wheat) can, conceptually speaking, not be reduced to its mere physical presence: it is always also part of a (Western) property and social system and thus interwoven with specific conditions of production, exchange, consumption and reproduction. Hence, alongside economic transactions and the spatial and temporal flow of a commodity, local livelihoods and their resource structures are shaped and, vice versa, are structuring commodity chains. 'A livelihood comprises the capabilities, assets (including both material and social resources) and activities required for a means of living' (Scoones 1998: 5). Livelihoods are thus the very locus where the ability to produce or to buy food intersects with income, health and nutritional status of the different members of a reproduction unit – as the family in the opening paragraphs reveals. Thus, transnational corporations as well as local livelihoods have to be investigated in order to understand comprehensively the risks in global food chains.

As the resource structure delineates livelihoods to a great extent, it will further be argued that long-term stress, particularly on incorporated resources (i.e. on health and nutritional status), is inscribing bodies. Inscribed bodies – the poor, hungry and starving, but also the overweight – are ultimately to be considered as an integral part of cross-continental food chains. Hence, the chapter stresses the need to link agri-food-complex and commodity-chain approaches with concepts of vulnerability and embodied uneven development.

From the analytical perspective, a food system is delineated as a socioeconomic and spatial system – comprising different levels of society – that can be divided into three overlapping and interacting subsystems (Figure 8.2): global agricultural food production; (urban) market exchange; and food consumption and social reproduction.[2] The *potential of exposure* to risks within a food system subsequently depends on the conditions within the realms of production, exchange and social reproduction, while the *coping capacity*, delineating the possibilities of buffering these risks, is a consequence of the demographic composition of the reproduction unit (i.e. household) and the access of its members to resources.[3] Hence, access, resources and social reproduction need to be explained in more detail. Access means the 'ability to derive benefits from things' (Ribot and Peluso 2003). In contrast, the notion of resources – transformed from the ancient idea of reciprocal and regenerative relationships between humans and nature (Shiva 1999) into a utilitarian concept of 'inputs' for livelihoods – refers here, more openly, to the capability of doing things. Drawing on the work of Bourdieu (1983) and Giddens (1995) on 'capital' and 'allocative resources', four forms

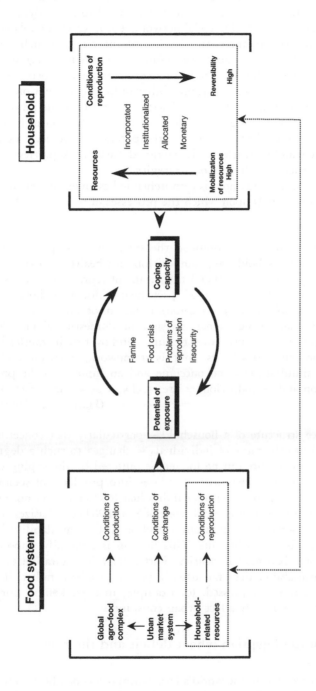

Figure 8.2 Analytical perspectives: risks of vulnerability towards food insecurity

Source: Author.

of resources are distinguished (Gertel 2002b): (1) incorporated resources (i.e. nutritional and health status etc.), which are bound to the body; (2) socially institutionalized resources (i.e. social networks etc.), which are related to the subject; (3) allocative resources (i.e. land, machinery etc.), which are linked to property rights; and (4) monetary resources (i.e. savings etc.), which also depend on property rights, but can be more easily exchanged between people. This concept of resources offers the possibility of linking the notion of uneven development with property rights and with the status of the physical body of a person, providing insights into the social reproduction of households. The mutual convertibility of resources, the potentiality to accumulate only certain resources and their selectively restricted use by third parties are crucial mechanisms structuring livelihood strategies.

In this respect, the household is comprehended as an empirical and analytical 'unit' (Wong 1984) upon which the processes of social reproduction are rooted:

> Forces of reproduction comprise the quantity and quality of labour (affected by household size, composition and health status), the means of consumption (e.g. food) and the means of reproductive work (water, housing, technologies for food preparation, for household maintenance and for biological reproduction). Relations of reproduction comprise the gender and generational division of domestic labour, decision-making power and control over resources and over their transfer between generations and households. In this framework, nutrition and health status is simultaneously an outcome and an input into the process of production and reproduction, as are stocks of goods and of money.
>
> (Harris *et al.* 1990: 2784)

If the resource structure of a household – particularly that concerning the nutrition and health status of individuals – changes to such a degree that the potential of exposure can no longer be buffered by the coping capacity of the household, then insecurity translates into problems of social reproduction, and even the occurrence of individual food crises can no longer be excluded (Figure 8.2). The seminal work of Sen (1981) emphasizes that food insecurity may not only result from a decline in food availability but also from declining food entitlements. This finding is especially important for the analysis of urban food insecurity. Particularly in capital cities such as Cairo, food availability does not usually pose a problem; rather, it is food entitlement decline – expressed, for example, in a weakening purchasing power – that can seriously affect urban consumers.

Production: the Egyptian wheat deficit and the role of US imports

Egypt is, as noted, one of the world's largest importers of wheat.[4] The degree and structure of the Egyptian import dependency is most visible in the national wheat deficit (Figure 8.1). While national wheat production did not change considerably until the mid-1980s, wheat imports accelerated in the early 1970s, supported by the so called 'open door' policy. National production started to increase in the mid-1980s, enabled by the expansion of agricultural areas and due to higher yields per area. However, a considerable amount of the locally produced wheat is consumed directly by the Egyptian producers and does not leave the rural areas (Hopkins *et al.* 1995). In the financial year 2002–3, for example, only about 18 per cent of the required total consumption of wheat was purchased locally in order to produce bread (Foreign Agricultural Service of the US Department of Agriculture (FAS) 2003b). Imports are thus predominantly provisioning the large cities – first of all, Cairo.

Egypt's wheat-import dependency also reflects the structure of the global agri-food system (Goodman and Watts 1997). Today, wheat for export markets is predominantly produced in a small number of countries, such as the US, Canada, France, Australia, Argentina and recently Russia; these countries supply more than two-thirds of the total world exports (FAS 2003a). Postcolonial countries on the other hand are increasingly producing 'non-traditional' agricultural commodities, such as flowers, fruit and vegetables, provisioning the supermarkets of the rich 'North'. This spatial specialization is being reinforced along with the one-sided implementation of economic liberalization measures. While agricultural production in the US and EU is protected and subsidized, Egypt, for example, is forced to privatize its national economy and open up its markets for foreign agricultural products.

The US is playing a crucial role in the restructuring of Egypt's wheat provisioning system. America's relations with Egypt are characterized by deep asymmetries in the power structures during recent decades, and are metaphorically captured by Mitchell's term 'America's Egypt' (1995). Since it is no secret that goods and capital that are channelled to Egypt in the form of development aid flow back to the US in different ways. This kind of 'development cooperation' helps to maintain dependencies and to open up new markets for US agricultural products and other consumer goods. If one keeps the conflict in the Middle East in mind, it is also obvious that Egypt, after Israel, is the most important recipient of US aid for political reasons. The US Agency for International Development (USAID) is thus by far the most important donor for Egypt. Its (conditional) economic assistance is divided into different programmes, food aid being one of them. Under this sub-programme, Egypt received about US$3.8 billion between 1975 and 1996.

The Agricultural Trade Development and Assistance Act of 1954 (known as PL-480), became, however, the most important instrument of US food aid. PL-480 was initially intended to support low prices for expensive US surplus production. Later in the 1960s its use was connected with military and security goals. Starting from the 1970s when international grain prices increased sharply, the original aid character was replaced by a market-oriented export function (Clay and Shaw 1993), while Egypt advanced to become its largest recipient worldwide. When, in 1993, Egypt ultimately stopped buying under PL-480, US interest groups – such as the wheat councils – had already succeeded in making Egypt an important market for commercial wheat sales. They did so with the support of the United States Department for Agriculture (USDA), and by using indirect subsidization programmes, like General Sales Manager (GSM-102) and the Export Enhancement Program (EEP) (Gertel 1998). In 1995 about 95 per cent of all Egyptian wheat imports originated from the US, declining only to about 82 per cent in 2000, and realizing export earnings of about US$0.54 billion (FAS 2003b). Supplying wheat for Egypt's poor is obviously a profitable market, and US food aid prepared the taste for (commercial) wheat imports over the past 30 years. However, an investigation into the risks of local food insecurity requires the inclusion not only of (public) policies but also of (private) company strategies.

Marketing: TNCs as global players in the grain trade

As early as the 1970s it was apparent that the market power of large corporations was vital to the organization of the international grain trade (Morgan 1979). The so-called 'big five' of the 1970s – Cargill, Continental Grain, Louis Dreyfus, Bunge y Born, and Cook Industries – have conducted, at various times, over 90 per cent of the US and over 80 per cent of the Argentinian wheat-export trade, and Cargill and Continental Grain alone have also controlled 90 per cent of the grain exports of the European Union (Gowers 1986). In the last few years, mergers further accelerated the concentration of the grain business: by 2000, Cargill, Archer Daniels Midland (ADM) and Zen Noh handled 81 per cent of the US corn exports and 65 per cent of soybean exports. Cargill and ADM are also among the top four in terminal grain handling and flour milling (Krebs 2003b). Already in 1998, Cargill took over the business of Continental Grain. The company then further extended its marketing scope by buying part of Australia's grain industry in 2002, when it acquired the flour milling assets of Goodman Fielder (Cargill 2002). Thus, Cargill now has come to be considered 'the' global player in the world grain business. Frank Sims, president of Cargill's North American grain division, comments: 'Cargill has become an ardent proponent of eliminating . . . government policies that artificially skew the production, consumption and trade of agricultural products. Consequently, the company favours policies that promote open competition and allow markets to work well' (Hoy 2002).

Exact details about the real market power of these corporations are, however, difficult to obtain, because the companies are often family firms that operate within closed social networks (Heffernan and Constance 1994) and do not fall under disclosure rules that many governments enforce. For example, Continental Grain, which was owned by the multimillionaire Michael Fribourg, did not publish a single balance sheet until the mid-1980s (Gowers 1986). Other 'big five' companies are also private organizations structured by family ownership. Cargill, for example, is the largest privately held firm in the US. Being over 130 years in business, still 83 per cent of the company is owned by the MacMillan and Cargill families (Weinberg and Copple 2002). It operates in 61 countries, has more than 800 offices, employs about 98,000 people, and its revenues amounted to over US$60 billion in 2002 (Forbes 2003). From the time of its inception until the early 1990s, the company only employed five managing directors (Broehl 1998).

The expansion of 'corporation empires', predominantly that of Cargill and (prior to its being taken over) Continental Grain, is closely linked with their access to US government institutions and programmes (Kneen 2003). The exploitation of US food-aid programmes started in the 1960s: as an example, in 1963 US agricultural exports reached a total of about US$4 billion. The US government alone provided about US$1.5 billion to finance food aid. Cargill, and other companies that carried out this exchange, realized tremendous profits. Krebs states: 'By this time, PL-480 had already generated US$2 billion in sales for Continental and Cargill alone' (1992: 329). In Egypt the TNCs divided the most important market shares. Under the PL-480 (Title I) programme, wheat sales during the four years (inclusive) 1975 to 1978 amounted to US$99 million for Louis Dreyfus, US$85 million for Bunge, US$84 million for Cargill and US$73 million for Continental Grain (Gilmore 1982: 264). During the period of controversial grain sales to the Soviet Union (1972 to 1974), the corporations were also successful in getting a large share of the lucrative orders (Porter 1984).

This was enabled not least through a web of interests that existed between the TNCs and government institutions, such as the USDA, where previous top leaders of the large companies held important positions (Krebs 1992: 335). In 2003, Warren R. Staley, Cargill's Chairman and Chief Executive Officer, was appointed by President Bush to serve on the President's Export Council, the premier national advisory committee on international trade (Krebs 2003a).

Provisioning: privatizing public food supply

This concentration of market power within the wheat commodity chain has a crucial impact on local food security in Cairo, even more so because it coincides with the period of structural adjustment during which the Egyptian government has had to relinquish its buffer function, which used to protect low-income consumers. Economic restructuring contains, for example, the release of currency rates and interest rates, the lifting of price controls for agricultural and industrial products, further limitations on state subsidization, the reduction of import and export limitations, and new legislative measures for the promotion of private investment, as well as the privatization of state-owned enterprises. In respect to the restructuring of the wheat provisioning system, two processes have been crucial: the privatization of the wheat processing industry and the cuts in the public food subsidy system.

The story of deregulating the national wheat industry goes as follows. In 1990, private companies were allowed to import wheat flour for the first time. Two years later private wheat imports were approved by the Egyptian government. Then, in the mid-1990s, the ownership of the public grain mills was partly privatized, while the private sector was encouraged to invest in new mills. By 2003, 36 private mills, mostly controlled by a few US companies, were operating. They processed 23 per cent of the total national consumption, but were limited to the so-called '72-per-cent flour' sector (used in expensive bread and pasta). On the other hand, from the 109 public mills that are allowed to produce '82-per-cent flour' (the most important ingredient for the popular and still subsidized *baladi* bread), the majority of the shares are sold to the private sector. Until today the public Food Industries Holding Company in Egypt still maintains political control over these mills. However, anticipating the trajectory of privatization measures, the Egyptian government is structurally preparing the way to abandon an important segment of its ability to control the strategic wheat food chain, rendering its population more vulnerable to entitlement failures.

Parallel to this, Egypt's food subsidy system is subject to major restructuring. From 1980–1 to 1996–7, public spending was dramatically reduced from 14 per cent of government expenditure to 5.6 per cent. Subsequently, only four foods remain subsidized: *baladi* bread, wheat flour, sugar and cooking oil – amounting to 3.74 billion Egyptian pounds (US$1.1 billion) in 1996–7 (Ahmed *et al.* 2001). Bread and wheat flour alone received 77 per cent of the subsidies, indicating its strategic significance – expressed locally in an allegedly stable price for bread at 5 piaster per loaf. In real terms, retail prices are, however, already increasing: loaves are losing weight, and the quality of the bread is decreasing by milling the wheat into flour at a higher extraction rate. In 1990, for example, the extraction rate for the production of *baladi* bread was increased to 82 per cent; hence the quality decreased. One year later the weight of a *baladi* loaf was reduced. Starting in the late 1990s, the taste of the bread was also altered, by adding (cheap) white maize as a new ingredient. Price increases in the form of weight and quality reduction may not be easily visible, particularly as the related changes are not implemented at once, they are also not implemented everywhere to an equal extent, and they have sometimes been coupled with selective wage increases. This strategy of 'gradualism' – slow, indirect price increases – avoids drastic price effects, such as those triggering the 'bread riots' in 1977. But although the underlying complex causal structure may be unknown to most consumers, the net result is obvious: *aysh* – bread and life – is becoming increasingly expensive.

Consumption: inscribing the bodies of the poor

The structure of food consumption at the household level in Cairo is determined by numerous factors: by the religious calendar, by eating habits, and, of course, by access to monetary resources. In this respect, the most vulnerable households exposed to food insecurity are those who rely heavily on discontinuous income by selling their labour power and simultaneously have little or no command over transfer income. The potential of exposure further increases for respective households if, for example, the main earner gets sick. The coping capacity in circumstances of poverty and insecurity is, however, often restricted to the mobilization of embodied resources. In these cases people such as the family in the opening quote have to accept changes in their nutritional and health status, caused, for example, by substituting meat or fruits with lower quality and cheaper foods such as bread, or by reducing expenditure for drugs and medical treatment.

Insecurity in Metropolitan Cairo manifests itself as complex socio-spatial fragmentation even within single urban quarters. Inequality is reflected in education (illiteracy rate), economic security (income), consumption (expenditure for bread), and in the stress on bodies (permanent sickness) (Table 8.1). Living conditions and housing type correspond closely with the dependency on bread consumption. Those who live in informal and rural situations have to spend high proportions of their income on bread. The same picture of insecurity holds true for the burden of morbidity: households in informal and in rural situations, and those living at the urban periphery, are forced to commit huge shares of their income on expenditures related to permanent sickness. Obviously, stress on monetary resources (i.e. high expenditure for food) translates into an exploitation of incorporated resources; hence people exposed to malnutrition may have difficulties to concentrate or even may die earlier. Moreover, bodily inscribed marginalization is not easily reversible. Even successfully targeted direct cash transfers, as an example for immediate intervention, would barely be able to re-establish a disenfranchized personal integrity, either physically or psychically.

In the absence of comparable empirical studies, the household survey from which the data in Table 8.1 is derived allows conclusions only on local food security for one point in time. However, the findings can also be situated in a longitudinal perspective: generally speaking, an increasing GINI-coefficient* indicates a widening social inequality alongside economic liberalization in urban Egypt during the 1990s (Adams 2000: 267). Already in 1995 about three-quarters (72 per cent) of the people living in informal housing situations in Cairo had to draw on allocated resources (assets, savings,

---

* The GINI-coefficient is an index commonly used to measure the inequality of a distribution of expenditure. It can be represented as $G = (2/nx) \, \text{cov}(y, r)$, where n is the number of observations, x is mean expenditure, y is the series total expenditure, and r is the series of corresponding ranks of expenditure.

Table 8.1 Inequality in Metropolitan Cairo, 1995

	Housing type			
	Private	Public	Informal	Rural
Old city core				
Income per capita (LE)	117.5	69.4	53.1	–
Illiteracy rate (%)	47.1	57.4	89.6	–
Bread: expenditure (%)*	12.2	16.9	53.3	–
Permanent sickness:				
expenditure (%)*	16.7	18.2	28.5	–
Established central area				
Income per capita (LE)	90.6	103.6	48.4	–
Illiteracy rate (%)	48.4	33.3	84.2	–
Bread: expenditure (%)*	8.9	8.5	62.5	–
Permanent sickness:				
expenditure (%)*	18.6	14.5	41.1	–
Young peri-urban area				
Income per capita (LE)	86.2	79.4	44.7	46.2
Illiteracy rate (%)	52.5	65.3	83.9	78.0
Bread: expenditure (%)*	11.8	10.6	41.7	26.0
Permanent sickness:				
expenditure (%)*	13.6	24.0	22.9	18.9
Well-to-do comparative area				
Income per capita (LE)	410.6	152.9	38.5	46.7
Illiteracy rate (%)	5.7	14.0	75.5	77.6
Bread: expenditure (%)*	3.4	4.9	34.9	22.7
Permanent sickness:				
expenditure (%)*	9.7	14.8	33.5	28.7
All				
Income per capita (LE)	176.2	101.3	46.2	46.5
Illiteracy rate (%)	38.4	42.5	83.3	77.8
Bread: expenditure (%)*	9.1	10.2	48.1	24.4
Permanent sickness:				
expenditure (%)*	14.7	17.9	31.5	23.8

Source: Gertel (2002a: 34).[5]

Note: The asterisk (*) indicates that the reference here is to stable income – that part of the overall income that is regularly available within a specific period without fluctuations.

debts) in order to buffer food price increases. And it was predominantly the people from the lowest income quartile (38 per cent of all households) who had to extend bread consumption during the previous three years – further substituting it for more expensive food – in order to buffer entitlement declines.[6] Poor bodies in Cairo are, thus, inscribed by the consumption of US wheat, and simultaneously are also profitable for the life sciences industry; both markets – for wheat and drugs – are structured by a few multinational companies.

Conclusion: accountability in cross-continental food systems

Within cross-continental food systems the growing demand for wheat in postcolonial countries coincides with economic liberalization and increasing social inequality. Here, urban areas in particular are at risk, as the urban poor – growing into a 'hungry' market over recent decades – are extremely dependent upon a secure supply of (imported) wheat. This is reflected in the transformations of Cairo's food system. For Cairo, the greatest dependency is upon the US, whose food assistance and aid programmes prepared 'the taste' for wheat, which nowadays has to be satisfied via commercial imports. In such a liberalized environment the global wheat chain is thus shaped by private economic interests. Increasingly TNCs are regulating international agricultural markets and determining the living conditions of millions of people in Cairo and elsewhere by acting as price makers. Simultaneously, sovereign powers of the state are being transferred to a widely unknown 'private sector', which is operating on the international stage without sharing the interests of governments, such as national food security. These merchants are also not acting on the basis of long-term development plans linked to socio-political targets but rather on the basis of short-term profit margins, largely disconnected from social considerations. Thus, the term market failure reveals itself to be an euphemism for the everyday performance of markets, concealing within an allegedly neutral technical term the social forces at work in unequal exchange situations. Studying cross-continental commodity chains and international food provisioning systems thus requires a thorough investigation of their articulation with local livelihoods and the conditions of social reproduction. It is here where the social accountability that is necessary to evaluate exchange situations and food security has to be rooted.

Notes

1   The notion of a 'chain' invites, as an analytical device, a simplistic reading that amalgamates the distinct spheres of social networks (connecting, for example, producers with processors, etc.), the manifold ways of financial transactions and credit schemes (connecting, for example, electronic with human assessments of risks) and the specific – space and time-related – movement of a single commodity into an allegedly unified 'chain'. For the neglected role of households within this concept, see Dunaway (2001).

2   The model is simplistic in the sense that it is reducing complex social interactions into a countable number of key variables (such as household, resources etc.) and analytical interrelations between these variables (such as the potential of exposure vis-à-vis the coping capacity). However, in contrast to a linear reading prescription this model offers different entry points (production, exchange etc.) towards a comprehensive understanding of food security; it links the global and the local, and also addresses aspects of temporality, i.e. the reversibility of food insecurity. In this connection, food security exists 'when all people, at all times, have physical and economic access to sufficient, safe and nutritious food to meet their dietary needs and food preferences for an active and healthy life' (FAO 2004).

3   The emphasis is on resources, because it is the access to and the control over resources that create the prerequisite for action. However, Giddens rightly emphasizes that resources alone (these also include authoritative resources in his reading) do not comprehensively determine structure. Rather, structure is conceptualized as 'recursively organized sets of rules and resources' (1995: 25). He considers rules of social life 'as techniques or generalizable procedures applied in the enactment/reproduction of social practices' (1995: 21).

4   In 1998, for example, the top five world wheat importers were (in million metric tons): Egypt (7.4), Italy (7.1), Brazil (6.9), Japan (5.9) and Korea (4.7); in 2001 they were: Italy (7.5), Brazil (7.0), Iran (6.4), Japan (5.5), Algeria (4.5) and Egypt (4.4) (FAO 2003).

5   The survey in 1995 covered three locations (core, centre, periphery) within Metropolitan Cairo (Gamaliya, Rawd al Farag, Matariya) and one well-to-do comparative area (Muhandasin). Within each location, four housing situations (private, public, informal, rural) were distinguished. A rural background, however, was found in only two locations. The housing situation 'private' is the most common form in Cairo; social stratification here is very high. Information on illiteracy is related to the female head of household; information concerning permanent sickness relates to expenditure for drugs and medical consultations. The sample is not representative for Cairo (n = 704 households, 3,556 individuals).

6   In 1981–2 the poorest urban income quartile in Egypt derived 49 per cent of their calorie intake from wheat (Alderman and von Braun 1984: 31), while in 1997 the two poorest metropolitan income quintiles derived 56.3 per cent and 50.6 per cent respectively of their calorie intake from wheat (Ahmed *et al.* 2001: 68).

Bibliography

Adams, R.H. (2000) 'Evaluating the process of development in Egypt 1980–97', *International Journal of Middle East Studies*, 32: 255–75.

Ahmed, A.U., Bouis, H.E., Gutner, T. and Löfgren, H. (2001) *The Egyptian Food Subsidy System: Structure, Performance, and Options for Reform*. Washington (DC): International Food Policy Research Institute.

Alderman, H. and von Braun, J. (1984) *The Effects of the Egyptian Food Ration and Subsidy System on Income Distribution and Consumption*. Washington (DC): International Food Policy Research Institute.

Bourdieu, P. (1983) 'Ökonomisches Kapital, kulturelles Kapital, soziales Kapital'. In R. Kreckel (ed) *Soziale Ungleichheiten*. Göttingen Otto Schwartz, pp. 183–98.

Broehl, W.G. (1998) *Cargill. Going Global*. Hanover (NH): University Press of New England.

Cargill (2002) 'GrainCorp and Cargill selected as buyer of Goodman Fielder's flour mills', Media Release, 16 August 2002. Online: www.graincorp.com.au/docs/news/media Releases.html#2002.

Clay, J. and Shaw, E. (eds) (1993) *World Food Aid. Experiences of Recipients and Donors*. London: Villiers Publications.

Dunaway, W.A. (2001) 'The double register of history: situating the forgotten woman and her household in capitalist commodity chains', *Journal of World-Systems Research*, VII (1): 2–29.

Food Agricultural Organization (FAO) (2003) Faostat Database. Online: apps.fao.org/page/collections?subset=agriculture (accessed 1 November 2003).

Food Agricultural Organization (FAO) (2004) The Special Programme for Food Security. Online: www.fao.org/spfs (accessed 14 April 2004).

Forbes (2003) Cargill. Online: www.forbes.com/finance/lists/21/2003/LIR.jhtml?pass ListId=21&passYear=2003&passListType=Company&uniqueId=5ZUZ&datatype= Company> (accessed 23 January 2004).

Foreign Agricultural Service of the United States Department of Agriculture (FAS) (2003a) *Grain: World Markets and Trade*, Circular Series (FG05–03).

Foreign Agricultural Service of the United States Department of Agriculture (FAS) (2003b) *Egypt – Grain and Feed Annual*, Gain Report (EG3003).

Gereffi, G. and Korzeniewicz, M. (eds) (1994) *Commodity Chains and Global Capitalism*. Westport (CT): Praeger.

Gertel, J. (1998) 'La mondialisation du pain au Caire: profits des multinationales et insécurité locale', *Revue de Géographie de Lyon*, 73 (3): 219–25.

Gertel, J. (2002a) 'Globalisierung und Metropolisierung. Kairos neue Unsicherheiten', *Geographische Rundschau*, 54 (10): 32–9.

Gertel, J. (2002b) 'Globalisierung, Entankerung und Mobilität: Analytische Perspektiven einer gegenwartsbezogenen geographischen Nomadismusforschung', *Orientwissenschaftliche Hefte* (Orientwissenschaftliches Zentrum Halle: Mitteilungen des SFB: Differenz und Integration Heft 3), 57–96.

Giddens, A. (1992) *The Consequences of Modernity*. Stanford (CA): Stanford University Press.

Giddens, A. (1995 [1984]) *The Constitution of Society*. Cambridge: Polity Press.

Gilmore, R. (1982) *A Poor Harvest: The Clash of Policies and Interests in the Grain Trade*. New York: Longman.

Goodman, D. and Watts, M.J. (eds) (1997) *Globalising Food: Agrarian Questions and Global Restructuring*. London and New York: Routledge.

Gowers, A. (1986) 'How the merchants of grain are riding out the storm', *Financial Times*, 28 November: 40.

Granovetter, M. (1985) 'Economic action and social structure: the problem of embeddedness', *American Journal of Sociology*, 91 (3): 481–510.

Harris, B., Gillespie, S., and Pryer, J. (1990) 'Poverty and malnutrition at extremes of South Asian food systems', *Economic and Political Weekly*, 22 December: 2783–99.

Heffernan, W.D. and Constance, D.H. (1994) 'Transnational corporations and the globalization of the food system'. In A. Bonanno, L. Busch, W.H. Friedland *et al. From Columbus to ConAgra. The Globalization of Agriculture and Food*. Lawrence: University Press of Kansas, pp. 29–51.

Hopkins, N., Mehanna, S. and Abdelmaksoud, B. (1995) 'Farmers, merchants and primary agricultural marketing in Egypt'. In J. Gertel (ed.) *The Metropolitan Food System of Cairo*. Saarbrücken: Verlag für Entwicklungspolitik, pp. 43–67.

Hoy, A. (2002) 'Grain reaction', *The Bulletin*, October 16 (with Newsweek): 1–2.

Institute of National Planning (INP) (1996) *Egypt Human Development Report 1996*. Cairo: Institute of National Planning.

Kneen, B. (2003) *Invisible Giant: Cargill and its Transnational Strategies*. London: Pluto Press.

Korayem, K. (1996) *Structural Adjustment, Stabilization Policies, and the Poor in Egypt*. Cairo: The American University Press in Cairo.

Krebs, A.V. (1992) *The Corporate Reapers: The Book of Agribusiness*. Washington (DC): Essential Books.

Krebs, A.V. (2003a) 'Cargill Chief Executive named to President's National Advisory Committee on International Trade', *The Agribusiness Examiner*, 277, 7 March. Online: www.ea1.com/CARP (accessed 23 January 2004).

Krebs, A.V. (2003b) 'Hendrickson-Heffernan study reveals high degree of concentration in U.S. agricultural markets', *The Agribusiness Examiner*, 144, February 27. Online: www.ea1.com/CARP (accessed 23 January 2004).

Mitchell, T. (1995) 'The object of development. America's Egypt'. In J. Crush (ed.) *Power of Development*. London: Routledge, pp. 129–57.

Morgan, D. (1979) *Merchants of Grain*. New York: Viking Press.

Porter, R.B. (1984) *The U.S.–U.S.S.R. Grain Agreement*. Cambridge: Cambridge University Press.

Raikes, P., Jensen, M.F. and Ponte, S. (2000) 'Global commodity chain analysis and the French *filière* approach: comparison and critique', *Economy and Society*, 29 (3): 390–417.

Ribot, J.C. and Peluso, N.L. (2003) 'A theory of access', *Rural Sociology*, 68 (2): 153–81.

Scoones, I. (1998) 'Sustainable rural livelihoods: a framework for analysis', *IDS Working Paper* 27. Brighton: Institute of Development Studies.

Sen, A. (1981) *Poverty and Famines: An Essay on Entitlements and Deprivation*. Oxford: Clarendon Press.

Shiva, V. (1999) 'Resources'. In W. Sachs (ed.) *The Development Dictionary: A Guide to Knowledge as Power*. New York: Zed Books, pp. 206–18.

Watts, M.J. (1989) 'The agrarian question in Africa: debating the crisis', *Progress in Human Geography*, 13 (1): 1–41.

Weinberg, N. and Copple, B. (2002) 'Going against the grain'. Online: www.forbes.com/free_fobes/20002/1125/158.html (accessed 15 April 2004).

Wong, D. (1984) 'The limits of using the household as a unit of analysis'. In J. Smith, I. Wallerstein and H.D. Evers (eds) *Households and the World-Economy*. London: Sage Publications, pp. 56–63.

World Bank (1992) *World Development Report 1992*. Oxford: Oxford University Press.

9 The local cultures of contract farming

The export of fresh asparagus from the Philippines to Japan*

Sietze Vellema

Introduction

Industrialization and globalization of food provision in combination with increased attention to product quality and safety have fostered prominent and widespread institutional changes in the trade of fresh produce. These include the rise of contractual exchange in the place of (de-regulated) spot markets (Reardon and Barret 2000; Eaton and Shepherd 2001). This chapter examines the institutional dynamics of such a specific production arrangement, contract farming, and pays specific attention to the coordinating procedures and policing mechanisms resulting from the social and technical integration of independent farmers into a global agri-food system.

Contract farming of fresh fruits and vegetables links local communities to corporate strategic outlooks and competitive consumer markets. Accordingly, companies have to be knowledgeable about how to coordinate their activities and behaviour with the activities and behaviour of others. For understanding new forms of collaboration and institutional behaviour, the chapter focuses on the politics of institutional modalities linking corporate schemes and local communities, and for this purpose, it gets inside a Philippine contract farming scheme producing fresh asparagus for the Japanese market. Ethnographic research presented in this chapter identifies a pallet of institutional modalities, revealing the capacity of contractual arrangements to incorporate diverse social relations and institutional perspectives into a single organizational framework functional to production and marketing (Vellema 2002, 2003).[1]

With regard to cross-continental food chains, the key contribution this chapter makes is to give attention to the local dynamics implicit in international agri-food relations. This chapter perceives contract farming schemes

* The chapter is an output from the Programme of International Cooperation, the North–South Centre, Wageningen University and Research Centre, executed through a grant from the Netherlands Ministry of Agriculture, Nature and Food Quality. The Technology and Agrarian Development group, Wageningen University, sponsored fieldwork (from 1996 to 1998) and analysis.

as a locally embedded political coalition that engineers the political and organizational features of integration (cf. MacKenzie 1992). Such a perspective focuses attention on growers' agency and solidarity and explains the variety of relationships in contract farming (Little and Watts 1994). This approach is rooted in the observation that growers have different appraisals of how to act and, consequently, how to see the relation with and their dependence on international agri-business.

Theoretical perspectives

Many scholars in agrarian and rural sociology tend to concentrate on what goes on outside the 'globalizing corporation'. Agrarian sociology, in particular the new political economy of agriculture literature, examines contract farming through the lens of industrial restructuring (Friedland 1994), the new international division of labour in agriculture (Raynolds *et al.* 1993) and new patterns of trade and changes in retailing (McMichael 1992; Bonanno *et al.* 1994). Much attention is given to the coercive character of contracting (Gertler 1991) and to administrative hierarchies in international agribusiness (Rickson and Burch 1996). Actor-oriented studies in rural sociology alert us to the dangers of exaggerating the potency and driving force of external institutions and interests (Marsden *et al.* 1990, 1992; Long and Long 1992). They argue that, although agriculture is global in scope, each situation represents a specific configuration of interlocking actors' projects. Actor-oriented studies dedicate the resulting diversity mainly to the reaction of local networks of groups and associations to global conditions (Long and van der Ploeg 1994; Long 1996).

I conclude from the above that little has been said about what happens *inside* the organization and how specific organizational arrangements are deployed to enable companies to develop competitive advantage in, for example, the dynamic markets for horticultural products (Pritchard and Fagan 1999). Rather than explain the motivations for offering contracts in the context of globalization, I try to uncover how economic behaviour is embedded in a network of social relations, especially because independent farm-level decision-makers are key to ventures such as contract farming (Porter and Phillips-Howard 1997; Welsh 1997). My interest is in questions about the coordinating procedures and policing instruments inside the institutional configuration of contract farming schemes (Wolf *et al.* 2001; cf. Coombs *et al.* 1992). The analysis sees responses of contract farmers as collective as well as individual actions and locates their behaviour both inside the corporate structure and in the local community. Thus, elements of society crossing the boundaries of a corporate scheme constitute, together with a mixture of corporate management styles, the institutional modalities present in contract farming. Consequently, the analysis gives emphasis to the variety of social and political projects that steer performance of contract farming schemes.

I use neo-Durkheimian cultural theory as a heuristic device for mapping the institutional modalities connecting companies and growers. Neo-Durkheimian cultural theory, or grid-group theory (Douglas 1987, 1996), offers a straightforward framework to categorize organizational and social behaviour. Essentially, grid-group theory distinguishes four social formations or institutional orders as the social and cultural context of individual behaviour. These formations are constructed by measuring different types of individual and collective responses to incorporation (social involvement or group) and imposition (regulation or grid) (Thompson *et al.* 1990). The group and grid dimensions can be either strong or weak; for example, a hierarchical formation represents both high levels of integration and rule-based behaviour. In my analysis, behaviour of growers and company employees and managers is nested in institutional modalities, which combine different levels of regulation with modes of social involvement.

Contract farming in the Mindanao export fresh asparagus sector

Contract farming links farmers to downstream international agents (multinational firms, distributors, shippers, retailers), and ultimately to consumers. This case study examines the production of fresh, premium asparagus exported to Japan and contracted out by a subdivision of Dole Philippines. Dole's Philippine operations are mainly concentrated in Southern Mindanao (Figure 9.1) and involve the production of pineapples and bananas, and, recently, a diverse package of high-value crops largely marketed in Japan. In the early 1990s, softening prices for its major products in international markets – bananas and pineapples – and the Japanese appetite for fresh vegetables encouraged the company to venture into competitive fruit and vegetable markets. Globally, the Philippines is still a small producer of green asparagus, compared to the United States, Peru and Mexico (USAID-funded Asia Regional Agribusiness Project (RAP) 1995), but in the Japanese market the Philippines has become a major competitor for the leading producers. Crucial to the company's market strategy in Japan was to sell high-quality asparagus spears at prices significantly lower than its competitors. To bring this strategy into effect, a contract farming scheme was established, with the goal of ensuring growers' compliance with required quality and cost benchmarks.

The contract farming scheme was located near the company's existing infrastructure for pineapple and banana. Consequently, the company had to construct a way to incorporate the existing distribution of land into the corporate structure. In this case, the newly established contract farming scheme incorporated a variety of existing landholdings, thereby impacting upon the existing regional political economy of Southern Mindanao.[2] In the early twentieth century, Muslim communities largely occupied the region. Since the Second World War, settlers received land titles from

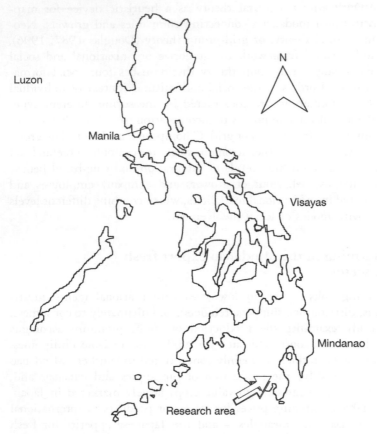

Figure 9.1 Mindanao, the Philippines
Source: Author.

the Philippines government and participated in an orchestrated development project. In the second half of twentieth century, the stream of migrants persisted, and eventually Christian settlers formed a majority in the region. In addition, the establishment of a vast pineapple plantation attracted large numbers of workers and exhausted the land frontier. These developments shaped the reality of land scarcity in which the company had to find its way; binding landowners through contracts was the option chosen.

In land-scarce situations, such as South-East Asia, contract farming schemes are responsive to existing patterns of land distribution and land reform legislation, inducing new ways to construct access to land and to develop new ways of organizing production (Vellema 2002; FAO 2002a, 2002b). In the Philippines, this has been facilitated through government policies favourable to agri-business development and financing schemes operated by semi-public, development-oriented banks.

By contracting out the actual production, the company gave up direct control over farm management and land. The role of contracts was to introduce predictability into agri-business operations, motivate performance and enhance quality control, and allocate financial risks and remuneration (Wolf *et al.* 2001). The contracts specified prices and quality requirements. Furthermore, the company exercised control at the point of production through advice and assistance from its technicians. However, typical contracts are incomplete; not all events can be anticipated and rationalized in a formal contract and the contract must function in a changing organizational and social environment (Rousseau 1995). Hence, ongoing and fluid interactions between management and local growers are inevitable in contract farming.

Local cultures in food chains

The observations above lead on to the question of what social processes accompanied the amalgamation of a variety of landowners, with correspondingly divergent moralities and political cultures, into a single scheme. In theory, all contract farmers, large or small, Muslim farmer or Christian settler, would have been subject to the same level of integration, as well as the same loss of control. However, the company adopted strategies that took into account variant local social formations, and combined these with various forms of corporate control. As it turned out, the prevalent assumption in much of the contract farming literature – that schemes operate as a uniform organization with a dominant culture – was found to be invalid. It then became crucial to comprehend the way in which social order is constituted in a particular institutional configuration.

Neo-Durkheimian cultural theory (see the previous section) provides a framework for this task. The ethnographic study of this case study revealed four major modalities in operation between the company and its contract growers (Figure 9.2). It is important to note that these modalities do not necessarily coincide with specific and bounded actor groups; growers could and did shift over time from one modality to another. Moreover, company management could and did employ different management styles in concert, responding to these institutional modalities. Hence, the modalities described in Figure 9.2 are not meant to categorize growers, but to pinpoint conflicting processes, arrangements and perceptions that encompass this contract farming arena. What follows is a brief review of key elements of each of the institutional modalities. For further reference, see Vellema (2002, 2003).

The first institutional modality is 'fatalist culture', which refers to a set of relations between the company and growers based around randomness and reciprocity. In this perspective, growers depended strongly on personal ties with company personnel and, consequently, reproduced traditional patronage relationships. The company's social obligation towards growers was most explicit in its relationship with persons in strategic positions, who acted as

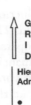

Fatalist culture Contrived randomness and reciprocity	Hierarchist culture Administrative control and transparency
• Acceptance of uncertain procedures and unpredictable rewards. • Stratified individuals alive at margins of organizational patterns. • Individual bargaining outside formal reward system and exploit personalised relationships in situation of limited choice. • Permanently failing organizations and perception that economic survival is not linked to performance. • Clash of cultures; close territories and put blame on the system. • Minimum anticipation and lack of disposition to take responsibility: ad hoc responses to events.	• Elaborate organizational apparatus of controllers and overseers at all levels of the organization; greater managerial grip. • Division of labour and differentiated roles. confidence in organizational competence. • Excessive trust in technical expertise and tight bureaucratic procedures. • Binding prescription and inability to learn; error inducing organization. • Fairness consists of equality before the law; conceal evidence of failure.
Individualist culture Entrepreneurial partnership and performance by competitive individuals	Isolationist culture GROUP Brokerage and delegated negotiation
• Freedom to enter and to exit transactions. • Relationship is subject to negotiation and dependent on the ability of individuals to shape their work as they choose. • Boundaries are provisional and failure stems from lack of cooperation. • Unchecked private gain at the expense of what is supposed to be a collective enterprise. • Pursuit of personal rewards; dependent on effort.	• Sending a delegate as negotiator; strong relations between group members. • Shared opposition to outside world keeps group bound together. • Respond collectively to opportunities provided by company. • Failing negotiation and unwillingness to accept higher authority to break deadlocks; lack of ability to resolve disputes and feuds.

Figure 9.2 Dimensions of and responses to incorporation and imposition in contract farming

Source: Fieldwork observations and analysis (Vellema 2002); drawing on Mars (1982), Douglas (1987), Hood (1996, 2000).

Note: The vertical *grid-axis* represents the degree to which life and behaviour of individuals are circumscribed by conventions and rules or by externally imposed prescriptions; it reflects the extent to which space for individual negotiations is reduced. The horizontal *group-axis* represents the degree to which individual choice is constrained by group choice; it reflects the extent to which an individual's life is circumscribed by the notion of solidarity of the group he or she belongs to.

intermediaries and fixed social problems. These persons were often able to turn brokerage into an economic-cum-political activity (cf. McCoy 1994). In the beginning of the scheme, the company had to persuade growers to sign a contract by showing that its offer was promissory. Under such conditions, loyalty was secured by provision of credit and strategic uses of advances and other payments (Clapp 1988). More ambivalent was the company's attitude towards cultural expressions of reciprocity in the case of individual growers. Individual growers settled for small concessions, such as non-operational advances or specific attention to production problems, but they had to deal with numerous gatekeepers inside the company's structure.

To achieve some security through interdependence, growers cultivated a personal network of alliances (Pertierra 1997). In addition, company employees actively nourished the idea of a personal partnership to sustain this level of confidence. Hence, signing a contract suggested the existence of new forms of reciprocity (cf. Hollnsteiner 1973). This perspective was successful initially in constructing trust between some growers and the company; however over time, reciprocal relations broke down when the company was faced with rising labour costs and declining productivity and quality levels.

Managerial interventions resulted from rising costs and declining quality levels in the scheme, which induced a stronger emphasis on the second institutional modality: 'hierarchist culture'. This modality refers to a set of relations between the company and growers based around the rigid application of legalistic and technical procedures. The use of financial figures and accounting procedures as important tools for reviewing performance and reorganizing relationships in the scheme, presupposed an application of universal principles (Jönssen 1996). The company's review process was non-discriminatory: big landlords as well as small farmers were threatened with the same treatment. This process confronted growers with impersonal institutions and objective financial norms enforced by remotely known functionaries. Increasingly, management rewarded efficiency and placed less value on personal ties. Impersonal performance reviews seriously undermined existing vertical links binding growers to office-holders or technicians, which had raised expectations in regard to services and security (cf. Scott 1972). Obviously, the performance review left no space for showing gratitude or consideration for growers who had played a stimulating role in the early phases of the venture.

Of overriding importance for the hierarchical modality was the fact that the company acted as collecting agent for a semi-public bank that provided loans and credit lines to growers' associations. Its role as bank agent turned the company's financial department into a key player in transactions between growers, associations, bank and company. In contrast to the work of the agricultural department, managers and financial staff had no direct involvement in agricultural production. For them, accounting information was an end to monitor the company's profitability. Financial figures were crucial in defending the outcome and the future of the operations in front of local management, representatives from headquarters and, in the end, company shareholders. Understandably, a typically instrumental interest in production dominated these spheres of the corporate structure (cf. Roberts and Scapens 1985).

The third institutional modality is 'individualist culture', which is based around entrepreneurial partnership and performance. In principle, the company perceived the relationship between growers and itself as a mutual dependency of two independent and equal business partners participating in a successful enterprise. Hence, company management took efforts in different ways to construct a 'partnership' with growers, who were considered business farmers who would negotiate individually for better terms and higher prices. However, the contradictory aspect of this institutional culture is that these partners clearly are unequal in size and power, and the company still has to find a way to construct control. Consequently, the 'individualist' culture combined with hierarchical measures and personalized relationships.

Growers operating within this modality tended not to solve problems with the company collectively. But at the same time, these 'entrepreneurial farmers' complained of an alleged 'lack of clarity' in their dealings with the company, and discredited the company's status as the technical expert in a risky enterprise. Additionally, growers objected to the accounting system that failed to create transparency in the numerous transactions between company, bank and grower. Although the company might not be bent on cheating growers as a matter of policy, errors in the accounting system eroded growers' confidence in the whole operation. Understandably, entrepreneurial farmers wanted to know not only that they were gaining financially from contract farming but also that they were not being cheated out of further profits by company manoeuvring (Jaffee 1994; Porter and Phillips-Howard 1997: 227).

Finally, there was an institutional modality that could be classified as 'isolationist culture', characterized by brokerage and delegated negotiation. In this modality, clashes over the validity of financial figures, misapplied accounting and unresolved technological puzzles forced the company to review its relationship with growers, and deal through intermediary brokers. These structures were particularly evident in stable Muslim communities, and involved a leading role of political leaders, or *datus*, as spokespersons for their constituencies. Muslim leaders explicitly nurtured ideas of justice and social equality; their sense of propriety was that people in their constituency, both growers and workers, were entitled to livelihood and dignity (cf. Hollnsteiner 1973; Wertheim 1978; Kerkvliet 1986). Typically, a diverse group of growers and workers belongs to the constituency of one single family, of which a member was assigned to negotiate on behalf of the group. Consequently, the brokers had to defend both their individual interests as well as collective interests in cases of failure or conflict.

For the company, this form of representation turned out to be burdensome. In the case of conflict, some brokers, for individual or collective reasons, decided to shift to an isolated position and obstruct further incorporation into the corporate structure. Disapproval of the company's formalized review process, sometimes threatening their persistence in the scheme, led to an isolationist position. Entire kinship networks became involved in such conflicts, affecting all growers in particular areas. Dealing with such complicated social matters was beyond the capacity of company technicians and officials. After several attempts to mediate, company management tended to leave the community to solve these problems; technicians were no longer allowed to enter the area and most operations were cancelled.

Conclusion

This chapter's reporting of an ethnographic analysis of the institutional modalities of contract farming arrangements in a developing world context reveals the fact that there is no institutional fix in cross-continental agri-food systems. It shows that global trade of fresh produce importantly depends on specific social and cultural conditions under which companies and producers collaborate to secure supply and to achieve quality (cf. Grossman 1998). This study's inquiry of contract farming centres on organizational variety and evolving definitions of management functions, as features of a system of vertical integration. It describes how an unspecified mixture of coercion and control, and of persuasion, conventions and converging self-interests, manufactures a variety of institutional modalities in contract farming.

The four institutional modalities of control and regulation, illustrated in Figure 9.2, acted as interdependent social practices receiving variable emphasis during the evolution of the fresh asparagus contract farming scheme in the Philippines. In these modalities, growers and company were held together differently. First, building personalized relationships constructed the idea of partnership in a new enterprise. Second, the administrative and financial inclusion of growers gave logic to an extensive corporate administrative hierarchy. Third, the process of review of performance also gave rise to subtle forms of competition and a move towards individualized bargaining. Finally, the brokerage of a social compromise led, in more extreme cases, to abandonment and obstruction. This chapter has described the messy mixture of institutional modalities as an unintended outcome of how contractual relations are embedded in local societies rather than as an intentional nurturing of different institutional cultures.

Company management utilized different techniques to respond to these four institutional modalities (Table 9.1), but eventually it became difficult to sustain such a mixture. Initially, the company was able to accommodate different modalities in a single organizational framework. However, external pressures, e.g. quality requirements in competitive markets, prompted a more strict managerial style addressing problems. Company managers hunted around among managerial styles, and disappointment over the capacity of one approach to deliver satisfactory results led to increasing support for one of the other options. Growers, of course, did not readily accept the organizational visions accompanying interventions in the scheme. Particularly, the increased emphasis on either hierarchical control or individualist performance became incommensurable with more personalized and culturally sensitive relationships.

Table 9.1 Managerial responses to institutional modalities in contract farming

Mutuality	Hierarchy
• Emphasize partnership and sustain familiarity	• Strengthen monitoring and performance review
• Reward local brokers for commitment	• Bring supervision and inspection in line with the ladder of authority
• Use reciprocity as solution to management problems	• Use formal power to pronounce on disputes or complaints
• Maintain unpredictable patterns of decision-making and supervision	• Command action and prescribe farming practices
Competition	**Brokerage**
• Make growers responsive to reward	• Broker social compromise and negotiate with collective
• Improve technical performance of individual farms	• Rely on group to check behaviour of individuals
• Outsource all activities	• Abandon area and elude negotiator
• Decentralize growers' association	
• Reduce involvement in productive activities	

Source: Adapted from Hood 2000.

The institutional approach developed in this chapter helps to understand social tensions in the management of contract farming. It describes how actual behaviour in the two domains, corporate schemes and local communities, becomes an, at times, fatally muddled compromise between potentially contending courses of action. It explains why certain forms of control collide with divergent social and cultural understandings of the nature of a contract.

Obviously, external dynamics resulting from market pressures, competitive strategies or corporate demands continuously interfere with the specific social balance found in a contract farming scheme. The conclusion of this chapter is that the capacity of companies to translate these external dynamics into a running organizational form determines the social robustness of a contract farming scheme. The analysis indicates that, in cases of pressure, managers may prefer to opt for congenial, prescriptive solutions and retreat into fixed technologies rather than more stressful and 'nitty-gritty' actions in difficult social interactions (Hood 2000). Such emphasis on technical and rational behaviour denies the social origin of coordination procedures and policing mechanisms in a division of labour (cf. Douglas 1987) and, eventually, may erode the social cohesion of contract farming.

Notes

1   The general approach used for this study was to develop an ethnography of contract farming. The approach particularly included qualitative methods: participant observation, semi-structured interviews, life and farm histories and situational analysis of particular instructive events (Vellema 2002). I investigated the following areas: input and prescription of technology, surveillance and monitoring of productive

activities, measurement of quality, and sharing and computation of revenues (cf. Wolf *et al.* 2001).

2   For more studies of the regional political economy in Southern Mindanao, see: Pelzer (1948); Ileto (1971); McCoy (1982); Beckett (1982, 1994); Hayami, Quisumbing and Adriano (1990); Muslim (1994); Tan and Wadi (1995); Azurin (1996).

References

Azurin, A.M. (1996) *Beyond the Cult of Dissidence in Southern Philippines and Wartorn Zones in the Global Village.* Quezon City: Center for Integrative and Development Studies, University of the Philippines Press.

Beckett, J. (1982) 'The defiant and the compliant: the datus of Maguindanao under colonial rule'. In A.W. McCoy and Ed.C. de Jesus (eds) *Philippine Social History: Global Trade and Local Transformations.* Quezon City: Ateneo de Manila University Press.

Beckett, J. (1994) 'Political families and family politics among the Muslim Maguinanao of Cotabato'. In A.W. McCoy (ed.) *An Anarchy of Families: State and Family in the Philippines.* Madison: The Center for Southeast Asian Studies, University of Wisconsin, pp. 285–309.

Bonanno, A., Busch, L. and Friedland, W.H. *et al.* (1994) *From Columbus to ConAgra: The Globalization of Agriculture and Food.* Lawrence (KS): University Press of Kansas.

Clapp, R.A.J. (1988) 'Representing reciprocity, reproducing domination: ideology and the labour process in Latin American contract farming', *Journal of Peasant Studies*, 16: 5–39.

Coombs, R., Saviotti, P. and Walsh, V. (1992) 'Technology and the firm: the convergence of economic and sociological approaches?'. In R. Coombs, P. Saviotti and V. Walsh (eds) *Technological Change and Company Strategies.* London: Academic Press.

Douglas, M. (1987) *How Institutions Think.* London: Routledge & Kegan Paul.

Douglas, M. (1996) *Thought Styles: Critical Essays on Good Taste.* London: Sage.

Eaton, C. and Shepherd, A.W. (2001) *Contract Farming: Partnerships for Growth.* Rome: FAO.

Food and Agriculture Organization (FAO) Ag 21 (2002a) 'Contract farming in Asia', *FAO AG 21 Spotlight magazine.* Online: www.fao.org/ag/magazine/spot2.htm (accessed 28 August 2002).

Food and Agriculture Organization (FAO) Ag 21 (2002b) 'Agribusiness and small farmers', *FAO AG 21 Spotlight magazine.* Online: www.fao.org/ag/magazine/0107sp.htm (accessed 28 August 2002).

Friedland, W. (1994) 'The new globalization: the case of fresh produce'. In A. Bonanno, L. Busch, W. Friedland, L. Gouveia and E. Mingioni (eds) *From Columbus to ConAgra.* Lawrence (KS): University of Kansas Press.

Gertler, M.E. (1991) 'The institutionalization of grower-processor relations in the vegetable industries of Ontario and New York'. In W.H. Friedland, L. Busch, F. Buttel and A. Rudy (eds) *Toward a New Political Economy of Agriculture.* Boulder (CO): Westview Press.

Grossman, L. (1998) *The Political Ecology of Bananas: Contract Farming, Peasants and Agrarian Change in the Eastern Caribbean.* Chapel Hill (NC) and London: The University of North Carolina Press.

Hayami, Y, Quisumbing M.A. and Adriano L.S. (1990) *Toward an Alternative Land Reform Paradigm: A Philippine Perspective.* Quezon City: Ateneo University Press.

Hollnsteiner, M.R. (1973) 'Reciprocity in lowland Philippines'. In F. Lynch and A. de Guzman (eds) *Four Readings on Philippine Values.* Quezon City: Ateneo de Manila University Press.

Hood, C. (1996) 'Control over bureaucracy: cultural theory and institutional variety', *Journal of Public Policy*, 15 (3): 207–30.

Hood, C. (2000) *The Art of the State: Culture, Rhetoric and Public Management*. Oxford: Oxford University Press.

Ileto, R. (1971) *Magindanao, 1860–1888: The Career of Datu Uto of Buayan*. Ithaca (NY): Department of Asian Studies, Cornell University.

Jaffee, S.M. (1994) 'Contract farming in the shadow of competitive markets: the experience of Kenyan horticulture'. In P.D. Little and M. J. Watts (eds) *Living Under Contract: Contract Farming and Agrarian Transformation in Sub-Saharan Africa*. Madison (WI): University of Wisconsin Press.

Jönsson, S. (1996) 'Decoupling hierarchy and accounting: an examination of trust and reputation'. In R. Munro and J. Mouritsen (eds) *Accountability: Power, Ethos and the Technology of Managing*. London: International Thompson Business Press.

Kerkvliet, B.J.T. (1986) 'Everyday resistance to injustice in a Philippine village', *Journal of Peasant Studies*, 13: 107–23.

Little, P.D. and Watts, M. (1994) 'Introduction'. In P.D. Little and M.J. Watts (eds) *Living Under Contract: Contract Farming and Agrarian Transformation in Sub-Saharan Africa*. Madison (WI): University of Wisconsin Press.

Long, N. (1996) 'Globalization and localization: new challenges to rural research'. In H. Moore (ed.) *The Changing Nature of Anthropological Knowledge*. London: Routledge.

Long, N. and Long, A. (1992) *Battlefields of Knowledge: The Interlocking of Theory and Practice in Social Research and Development*. London: Routledge.

Long, N. and van der Ploeg, J.D. (1994) 'Heterogeneity, actor and structure: towards a reconstitution of the concept of structure'. In D. Booth (ed.) *Rethinking Social Development: Research, Theory and Practice*. London: Longman.

McCoy, A.W. (1982) 'Introduction: the social history of an archipelago'. In A.W. McCoy and Ed.C. de Jesus (eds) *Philippine Social History: Global Trade and Local Transformations*. Quezon City: Ateneo de Manila University Press.

McCoy, A.W. (1994) '"An anarchy of families": the historiography of state and family in the Philippines'. In A.W. McCoy (ed.) *An Anarchy of Families: State and Family in the Philippines*. Quezon City: Ateneo de Manila University Press.

MacKenzie, D. (1992) 'Economic and sociological explanations of technical change'. In R. Coombs, P. Saviotti and V. Walsh (eds) *Technological Change and Company Strategies: Economic and Sociological Perspectives*. London: Academic Press.

McMichael, P. (1992) 'Tensions between national and international control of the world food order', *Sociological Perspectives*, 35: 343–65.

Mars, G. (1982) *Cheats at Work: An Anthropology of Workplace Crime*. London: Unwin Paperbacks.

Marsden, T., Lowe, P. and Whatmore, S. (eds) (1990) *Rural Restructuring: Global Processes and their Responses*. London: David Fulton.

Marsden, T., Lowe, P. and Whatmore, S. (1992) 'Introduction: Labour and locality: emerging research issues'. In T. Marsden, P. Lowe and S. Whatmore (eds), *Labour and Locality: Uneven Development and the Rural Labour Process*. London: David Fulton Publishers.

Muslim, M.A. (1994) *The Moro Armed Struggle in the Philippines: the Nonviolent Autonomy Alternative*. Malawi: Mindanao State University.

Pelzer, K.J. (1948) *Pioneer Settlement in the Asiatic Tropics: Studies in Land Utilization and Agricultural Colonization in Southeastern Asia*. New York: American Geographical Society.

Pertierra, R. (1997) *Exploration in Social Theory and Philippine Ethnography*. Quezon City: University of the Philippines Press.

Porter, G. and Phillips-Howard, K. (1997) 'Comparing contracts: an evaluation of contract farming schemes in Africa', *World Development*, 25: 227–38.

Pritchard, B. and Fagan, R. (1999) 'Circuits of capital and transnational corporate spatial behaviour: Nestlé in Southeast Asia', *International Journal of Sociology of Agriculture and Food*, 8: 3–20.

Raynolds, L., Myhre, D., McMichael, P., Figueroa, V. and Buttel, F.H. (1993) 'The "new" internationalization of agriculture; a reformulation', *World Development*, 21: 1101–21.

Reardon, T. and Barret, Ch.B. (2000) 'Agroindustrialization, globalization, and international development: an overview of issues, patterns and determinants', *Agricultural Economics*, 23: 195–205.

Rickson, R.E. and Burch, D. (1996) 'Contract farming in organizational agriculture: the effects upon farmers and the environment'. In D. Burch, R.E. Rickson and G. Lawrence (eds) *Globalization and Agri-food Restructuring: Perspectives from the Australasia Region*. Aldershot: Avebury.

Roberts, J. and Scapens, R. (1985) 'Accounting systems and systems of accountability – understanding accounting practices in their organisational contexts', *Accounting, Organizations and Society*, 10: 443–56.

Rousseau, D.M. (1995) *Psychological Contracts in Organizations: Understanding Written and Unwritten Agreements*. London: Sage.

Scott, J.C. (1972) 'The erosion of patron–client bonds and social change in rural Southeast Asia', *Journal of Asian Studies*, 22: 5–37.

Tan, S.K. (1995) *The Socioeconomic Dimension of Moro Secessionism*. Quezon City: City for Integrative and Development Studies, University of the Philippines (Mindaneo Studies Report no. 1).

Tan, S.K. and Wadi, J. (1995) *Islam in the Philippines*. Quezon City: Center for Integrative and Development Studies, University of the Philippines (Mindanao Studies Report no. 3).

Thompson, M., Ellis, R. and Wildavsky, A. (1990) *Cultural Theory*, Boulder (CO): Westview Press.

USAID-funded Regional Agribusiness Project (RAP) (1995) 'World market for asparagus', *Rap market information bulletin*, no. 2. Online: www.fintract.com/rap/bulletins/market/aspar.html (accessed 31 January 1996).

Vellema, S. (2002) *Making Contract Farming Work? Society and Technology in Philippine Transnational Agribusiness*. Maastricht: Shaker Publishing.

Vellema, S. (2003) 'Management and performance in contract farming: the case of quality asparagus from the Philippines'. In S. Vellema and D. Boselie (eds) *Cooperation and Competence in Global Food Chains: Perspectives on Food Quality and Safety*. Maastricht: Shaker Publishing.

Welsh, R. (1997) 'Vertical coordination, producer response, and the locus of control over agricultural production decisions', *Rural Sociology*, 62: 491–507.

Wertheim, W.F. (1978) 'De Indonesische Moslims: meerderheid met minderheidsmentaliteit (chapter 11)'. In W.F. Wertheim, *Indonesië: van Vorstenrijk tot Neo-kolonie*. Meppel: Boom.

Wolf, S., Hueth, B. and Ligon, E. (2001) 'Policing mechanisms in agricultural contracts', *Rural Sociology*, 66: 359–81.

Part III

Lead firms and the organization of cross-continental food chains

10 Responsible retailers?

Ethical trade and the strategic re-regulation of cross-continental food supply chains*

Alex Hughes

Introduction

Since the mid-1990s, ethical trading initiatives have promised to reduce some of the economic and social disparities produced through exploitative cross-continental food supply chains. In the context of a trading landscape underpinned by neo-liberal rationalities and agendas, systems of provision for food consumed in advanced capitalist economies have increasingly involved global supply chains and low-wage production in economically less developed countries (Watts 1996; Goodman and Watts 1997). Academic studies taking a commodity chain approach have highlighted the social and economic inequalities created by supply relationships between Northern buyers and Southern producers (Cook 1994, 1995; Hartwick 1998).[1] Both the media and civil society organizations (CSOs) have also drawn public attention to the contrast between the consumption of branded goods sold at high prices in retail stores and the poor conditions of work at sites of export production. In the US, journalistic exposés and political campaigns have predominantly targeted brand manufacturers and retailers in the garment sector, for example the Stop Sweatshops Campaign (Johns and Vural 2000; Adams 2002). Elsewhere, including the UK, food retailers have also come under the critical spotlight (Hughes 2001a, 2001b; Freidberg 2003). As a response to media-generated public concern about labour conditions at sites of production, a large number of high-profile retailers and brand manufacturers have embarked upon strategies of ethical trade (Hughes 2001b; Jenkins 2002; Roberts 2003). Ethical trading initiatives stand in marked contrast to projects of fair trade, the latter involving more developmental objectives of empowering producers through 'alternative' supply chains

* I would like to thank Niels Fold, Bill Pritchard and two anonymous referees for constructive comments on an earlier draft of this chapter. I am very grateful to The British Academy for funding research on corporate learning in ethical trade (Award Number SG-33442). Finally, I would also like to acknowledge the corporate interviewees, without whom the research would not have been possible.

(Whatmore and Thorne 1997; Raynolds 2000, 2002; Renard 2003). Instead, ethical trade is understood by practitioners to be more of a corporate-led defence strategy against 'negative publicity' from pressure groups and the media (Blowfield 1999; Hughes 2001b; Adams 2002; Jenkins 2002; Roberts 2003). The starting point is almost always a code of conduct to ensure minimum standards for suppliers (Blowfield 1999).

The aim of this chapter is to evaluate retailers' strategic approaches to ethical trade, focusing on the specific case study of the UK food retailers. Most of the academic literature on ethical trade concerns its driving forces (Zadek 1998; Blowfield 1999; Jenkins 2002), its organization through multi-stakeholder approaches (Hughes 2001b; Blowfield 2002) and impacts on development (Barrientos *et al.* 1999; Barrientos 2000; Hale 2000; Hale and Shaw 2001; Hughes 2001a; Freidberg 2003). Engagement with corporate strategies for ethical trade is currently limited to descriptive accounts written by practitioners in the field (L. Roberts 2002; S. Roberts 2003). This chapter attempts to redress this imbalance in the literature by providing a critical discussion of managerial approaches to ethical trade. In particular, I want to highlight some of the limitations of ethical trading initiatives by demonstrating their embeddedness in the corporate decision-making processes and management systems of key retailers. Retailers' responses to the demands of ethical trade are continually negotiated through managerial reflexivity and corporate learning. However, perhaps unsurprisingly, the resulting strategies and organizational systems designed to deliver more egalitarian forms of supply chain management are frequently shaped as much by capitalism's imperatives as they are by a desire to improve labour conditions at sites of production. In attending to questions of corporate practice, this chapter is therefore part of a broader movement in agri-food studies to recognize the transformative capacity of social action by key agents in the food system (Arce and Marsden 1993; Goodman 2002), in this case the retailers. I argue that there have been three discernible waves of strategic development on the part of the UK food retailers, through which management systems for ethical auditing have evolved: (i) retail-led social auditing of own-label suppliers; (ii) third-party monitoring; and (iii) risk assessment and supplier self-evaluation. As retailers move sequentially through these phases of strategic development, it is suggested that their organizational approaches to ethical trade become progressively more aligned with commercial pressures than with a moral drive to produce ethical strategies in the name of the Southern worker.[2]

Ethical trade and the 'responsible retailer': the case of the UK supermarket chains

In the context of a *laissez-faire* national regulatory environment in which mergers and acquisitions and oligopolistic practice in the sector have rarely been challenged by the state, UK grocery retailing has been characterized by progressive concentration of capital since the 1950s. This consolidation of market power in UK food retailing really gathered pace in the 1980s amidst large-scale deregulation (Wrigley 1987, 1991). By the early 1990s, over 50 per cent of the total UK grocery market was constituted by just five retail chains (Hughes 1996). While the rank order of these chains and the competitive terrain has changed over the past decade, the concentration of capital in the industry has nonetheless continued. In 2002, the leading five supermarket chains accounted for 75.5 per cent of UK grocery sales (Institute of Grocery Distribution 2003). Such market power has translated into oligopsonistic buying power at the retailer–manufacturer interface (Grant 1987; Doel 1996), with the largest of the UK retail chains becoming increasingly able to dictate the terms and conditions of trade to their manufacturing suppliers. With minimal challenge from the regulatory state, these retailers extracted increasingly favourable pricing terms from their manufacturing suppliers, often involving non-cost-related discriminatory discounts. They extended their buying power still further by negotiating favourable terms of payment from these suppliers (Foord *et al.* 1992, 1996; Wrigley 1992; Bowlby and Foord 1995; Hughes 1996).[3]

Buying power exercised by UK retailers has not been confined to supply chains operating within the national boundaries of the UK. For the concerns of this chapter, it is important to note the growing significance of many economically less developed countries as producers of food retailers' own brands. Work by Barrett *et al.* (1999), for example, has drawn attention to the ways in which UK supermarket chains have forged increasingly direct relationships with suppliers of high-value horticultural produce from Kenya and The Gambia. However, research reveals the highly uneven power relations which characterize such cross-continental supply chains and which serve to perpetuate poor conditions of work at sites of production, including a lack of job security, pressure to work overtime at particular times of peak demand, and the employment of large numbers of temporary workers. Such a situation is compounded by low wage levels paid in economically less developed countries, the weak position of trade unions in many of these countries and an absence of tight regulations for the health and safety of workers (see Cook 1994, 1995; Hughes 2000).

The mid-1990s witnessed a surge of media-generated public concern over both environmental issues and worker welfare at sites of export production. Articles appeared in the UK broadsheets highlighting poor environmental and working conditions at production sites. Radio and television documentaries also became a part of the process with, for example, a BBC *Modern Times* documentary screened in 1997 revealing the means through which the UK supermarket chain Tesco sourced its own-label mangetout peas from Zimbabwe. In all cases, direct connections were made by the journalists and documentary film-makers between poor conditions at sites of production and the everyday purchase of the commodities through super-market chains. More direct pressure was at the same time exerted on retail corporations by CSO campaigns, with the most significant being Christian Aid's focus on supermarkets' global sourcing practices for their own-label food products (for a more detailed discussion of media exposés and CSO campaigns, see Freidberg 2004).

The key way in which the UK food retailers have responded to the recent public concern about their cross-continental supply chains has been their membership of the Ethical Trading Initiative (the ETI). Set up in the UK in 1997, this multi-stakeholder organization has been operating, at the time of writing, for just over six years as a civil initiative sponsored by the UK government's Department for International Development. By 2003, the organization was made up of 34 corporate members, 17 NGOs and repre-sentatives from four international trade unions (ETI 2003). One of the ETI's first tasks was to establish a code of labour conduct, which could be used by all corporate members to guide the implementation of responsible business standards in the context of their own supply chains. This Base Code consists of the following nine provisions, which build directly on core International Labour Organization (ILO) Conventions: (i) employment is freely chosen; (ii) freedom of association and the right to collective bargaining are respected; (iii) working conditions are safe and hygienic; (iv) child labour should not be used; (v) living wages are paid; (vi) working hours are not excessive; (vii) no discrimination is practised; (viii) regular employment is provided; (ix) no harsh or inhumane treatment is allowed (ETI 2000). The code applies to suppliers of the retailers' own-label products. Six of the seven retailers discussed in this chapter are ETI members and are therefore committed to this Base Code. The remaining company – Food Retailer E – has its own code, which includes the same conventions, with the addition of environmental clauses.

The first wave of strategic development: retail-led social auditing

The strategic commitment to the ETI and its Base Code, made by most of the UK supermarket chains in the mid to late 1990s, forms the basis of their corporate strategies for ethical trade. However, the *implementation* of the code in the context of individual retailers' supply chains remains under constant negotiation by each company. Since the inception of the ETI, retailers have continually sought to find the most effective ways of ethically monitoring their supply base. As Zadek (1998: 1427–8) has rightly pointed out: 'The fig leaves of codes of conduct are in themselves not enough . . . What is in addition demanded are reports of performance against these codes, externally verified.' Such reports are normally produced through social audits of pro-duction facilities. However, it should be noted that retailers are only just beginning to make significant progress in terms of ethically monitoring their supply base. In 2001, 7,989 producers supplying goods to ETI corporate members were ethically evaluated against the ETI Base Code, but this represents only 55 per cent of their aggregate, known supply base (ETI 2002: 25).

Each of the seven food retailers included in the study have a designated manager in charge of developing and overseeing ethical trade within the company. The practicalities of ethical trade are then dealt with by the technical teams – that is, by the food technologists whose jobs involve the management of product development, specifications and quality assurance. The main reason for this, cited by several interviewees, is that the technologists have traditionally had the most face-to-face contact with food suppliers. Moreover, since the 1990 Food Safety Act was passed, it has been the technologists on the part of the large retail chains who have conducted the food safety and quality audits at production facilities.[4] For these retailers, the most efficient way of organizing ethical trade initially was to bolt on social auditing to the technologists' job of technical monitoring, effectively conducting social auditing 'in-house'. This illustrates the way in which ethical trade extends retailers' regulatory control of the supply chain, which already involved food safety and quality audits (see Marsden *et al.* 2000). However, the flaws in this organizational strategy were soon recognized by the retailers themselves, as well as being criticized by CSOs, trade unions and the ETI. As the following two interviewees explain, the limitations of technologists' expertise prevent them from conducting thorough social audits with the majority of suppliers overseas – social audits that require particular skills involving worker interviews:

> To be an ethical auditor you've got to do ethical audits all the time and you can't be somebody who does technical audits most of the time and then . . . turns on your ethical auditing head and does that bit. It doesn't work. That's how we've approached these things in the past and so we come away thinking everything's fine when it isn't.
> (Ethical Trading Manager, Food Retailer A, 4 April 2003)

> We use a lot more external [auditors] and the amount of audits that the technologists are doing has gone down because obviously we realise there's limitations on what the technologists can do. You know, you can do basic health and safety, you can speak to management, but if you don't know the language, you don't *really* know the law, you can't interview the workers.
> (Ethical Trading Manager, Food Retailer B, 22 April 2003)

As the latter quote suggests, the strategic solution to the problem has been for retailers to contract out the auditing of their suppliers to specialist social auditing companies (Barrientos 2002). This move therefore characterizes the second wave of strategic development. While these independent, or 'third-party', audits do not completely replace the role played by the retailers' own technical teams, their use has nevertheless become a more prominent strategy for the ethical monitoring of food supply chains.[5] Moreover, third-party audits also provide retailers with the independent verification required to give credibility to their ethical trading programmes (Zadek 1998).

The second wave of strategic development: third-party social auditing

The largest of the specialist audit companies are international organizations with offices and employees located all over the world. The most frequently used audit firms, and the largest in terms of human resources and turnover, are Bureau Veritas, SGS and ITS. These companies also have their roots in technical, quality and safety audits. With the emergence of corporate social responsibility in the 1990s, though, they extended their function to include new methods of social and environmental auditing. Along with other smaller audit firms and selected management consultancies, the auditing expertise of these companies has been bought in by the UK food retailers and has consequently become a central part of the new economy developed in the name of ethical trade. Yet, far from simply using independent social auditing in an uncritical way, all of the retailers' ethical trading managers and their ethical consultants demonstrated an awareness of the shortcomings of such auditing as a method of ethical evaluation. Moreover, the broader-scale critique of auditing espoused by commentators such as Power (1997), Strathern (1997, 2000), Miller (1998, 2000) and Pentland (2000) is actually infiltrating the corporate mentalities of retail managers. Corporate inter-viewees identified a whole series of problems with social auditing. First, they discussed the problems inherent in the audit method itself:

> We have lots of examples of companies where they think they've had an ethical audit, but actually they've got problems that weren't picked up. The trouble is that any audit is rather like a snapshot in time, it's just a picture of a chess game in the middle and you can learn a lot about what's going on from that picture, but you can't learn it all, so that's one reason why audits are of limited value.
> (Ethical Trading Manager, Food Retailer A, 4 April 2003)

> The audit is a compliance monitoring tool, let's be very clear about that. And that's one part of the jigsaw. The other part of the jigsaw is raising people's awareness of standards and the importance of local solutions on the ground.
> (Head of Technical Policy and Strategy, Food Retailer C, 17 April 2003)

There is therefore an acute awareness on the part of the interviewees that audits are not some kind of unproblematic mirror on the realities of working conditions at sites of production. Rather, they recognize that many aspects of labour standards cannot be adequately evaluated through the audit process. Some managers and consultants also raised questions about the competences of the independent auditors who are under contract to monitor the implementation of the ETI Base Code on the retailers' behalf:

> I have serious concerns about some of [Audit Company A's] overseas auditing operations and I'm horrified that in Kenya some of the suppliers who have subsequently shown to be remiss were audited by [Audit Company C] or [Audit Company A] and it didn't pick up the issues that were actually out there.
>
> (Chairman, Responsible Sourcing Steering Committee,
> Food Retailer E, 13 May 2003)

Such worries are shared by the Social Compliance Manager at Food Retailer G, who emphasizes the role that her company has played in training the audit firms to update their skills in social auditing. There has also been the recent publication of a critique of independent labour monitoring, using the example of commercial audits conducted by PricewaterhouseCoopers (O'Rourke 2002).[6] A bias towards management at the factories and an inadequate use of worker interviews was found to be the crux of the problem.

Notwithstanding these concerns about the organizational limitations of social audits, it is arguably the price of hiring third-party auditors to monitor the supply chain which present the retailers with the most significant hurdle:

> We are paying [Audit Company A] over £1000 per audit. That is not an insignificant amount of money. And the reality is, it's not sustainable in any sort of sensible way forward . . . Just the cost of it, I mean, nobody can afford to do it. I mean, you have to be able to sell the product at the end of the day at a reasonable price . . . You know, you get to the stage where how much can you afford to spend? What is the most important thing about the food product – food safety or this ethical trade? I think I have to sit where I sit and say, well actually, it's safety.
>
> (Quality Assurance Director, Food Retailer D, 2 April 2003)

The audit price cited by the director above is generally acknowledged by ethical trading managers to be a realistic price charged by most auditing firms for an average one-day social audit of production facilities, though the length and cost of an audit varies marginally between companies and countries.[7] The prices charged by specialist auditing companies appear to add fuel to Retailer D's ambivalence towards ethical trade; an ambivalence that is effectively permitted by the neo-liberal context of private global regulation in which this director works.

The third wave of strategic development: risk assessment and supplier self-auditing

While third-party audits remain a central part of the monitoring systems for ethical trade, the aforementioned constraints mean that their use is limited by most food retailers. The most recent strategic manoeuvre on the part of the supermarket chains has been to use auditing in a more *targeted* way and to build new management systems to act as a guide to this targeting. To do this, ethical trading managers have effectively gone back to the original *commercial* motivation for responsible sourcing – the defence of the company against negative publicity. As S. Roberts (2003: 159) has noted, 'one of the key drivers for implementing CSR [corporate social responsibility] initiatives is a desire to avoid risks to corporate reputation'. This whole notion of risk to corporate reputation is driving the new management systems used by retailers for ethical monitoring. New techniques of risk assessment have therefore been developed by ethical trading managers and their consultants, illustrating how our 'risk society' (Beck 1992) and emergent 'risk consciousness' (Dannreuther and Lekhi 2000) has infiltrated practices of private interest regulation. These risk assessment approaches are being used to categorize own-label suppliers into high, medium and low risk groups. Otherwise known as the 'traffic light system' (red representing high risk, amber being medium and green signifying low risk), this approach is described below:

> We start off by risk-assessing for high, medium and low. We're not auditing the low risk, it's such a big job, you can't do it all. So we're primarily focusing on the high and medium. If we only had a short time, we'd get the high ones done first and then the medium. So the technologist goes through that . . . and we're now going through the phase where most of our high and medium have had external audits.
> (Ethical Trading Manager, Food Retailer B, 22 April 2003)

High-risk suppliers might therefore be those who are located in a country renowned for very poor labour conditions (including, for example, extremely low wages, human rights abuses and an absence of trade unionism). Particular sectors known for poor worker welfare can also be deemed high risk. In some cases, the retailers hire the expertise of consultants to aid them in this risk assessment exercise. Food Retailer C, for example, commissioned Consultancy B to produce a geographical and sectoral information 'toolkit', which helps their technologists to make an assessment about the risk posed by each of their suppliers. However, a supplementary means by which risk assessment is achieved involves the suppliers themselves completing questionnaires for the retailers:

> We have a risk assessment approach which is desk-based and it's based upon a number of criteria, the first of which is a questionnaire which

goes to a supplier . . . The questions ask about detailed issues relating to whether they comply with our code of conduct, which is ETI's Base Code . . . You look at the response to that and we might follow up a few issues with the supplier based upon that and then we also take into account what industry sector the supplier is in, where in the world it is and what we are aware of in terms of *potential* risks from contact with the ETI or NGOs or the trade union movement.

(Head of Technical Centre, Food Retailer F, 3 April 2003)

For most food retailers interviewed, this process helps them to target their limited resources for auditing towards the own-label suppliers who are likely to pose the greatest risk of negative publicity and subsequent damage to their retail brand. While representing a seemingly logical strategy in organizational terms, the degree to which this serves the needs of workers in export industries is highly questionable, as millions of labourers working for 'low risk' suppliers can quite literally fall out of account.

Hand in hand with techniques of risk assessment there are attempts to push the responsibility of monitoring back down the supply chain and on to the producers themselves. This is a strategy that has always been strongly favoured by Food Retailer G, and other supermarket chains have more recently followed suit. For Retailer G, the strategy of asking producers to conduct self-audits is fostered through what have been historically close, associative relationships with a small number of first-tier suppliers. This retailer not only asks these suppliers to conduct self-audits but also requires them to manage the monitoring of producers who in turn supply products to them. Independent auditors are simply brought in at the cost of the retailer to verify a sample of these audits and to audit any new producers brought into their supply chains. Otherwise, the retailer provides no further financial contribution to the costs of the self-audits. Interviewees also revealed that they had set up web-based management systems to facilitate this kind of supplier self-audit. With the help of Consultancy Company B, all but one of the UK food retailers interviewed (Retailer D) have very recently engaged in the project of Supplier Electronic Data Exchange (SEDEX). This scheme promotes the construction of a common on-line supplier database. Self-audit results from each participating supplier can be accessed through the use of a password by any retailer with whom they have a trading relationship.

Overall, such supplier self-auditing is argued by the retailers to provide organizational, financial and technical solutions to the challenges of ethical monitoring. But the retailers could be sharply criticized for evading a significant proportion of the costs and responsibility that arise from protecting their own corporate reputations. Moreover, within both the processes of risk assessment and on-line supplier self-auditing, the voices of the workers are largely silent. The management systems arising out of the third wave of strategic development for ethical monitoring therefore appear to embody even more management bias than the techniques used in previous strategies of audit, though third-party audits are designed to provide selected checks on this process. While there are some examples of more participatory approaches to ethical monitoring, whereby local CSOs and trade unions engage workers in longer-term and more developmental ethical evaluation,[8] these cases appear to be numerically dwarfed by the sheer volume of producers whose ethical assessment is captured more fleetingly by the modes of risk assessment and supplier self-audit described above.

Conclusion

While there is emerging opposition to the whole idea of having such voluntary initiatives for business responsibility, cross-continental food supply chains are nonetheless operating in a world trading environment in which the role of their regulation has effectively been handed to the private sector amidst progressive liberalization (Friedmann 1993; Raynolds et al. 1993; Watts 1996; Watts and Goodman 1997). This has encouraged the growth of private-sector solutions to ethical trading problems (Blowfield 1999; McClintock 1999; Tsogas 1999; Barrientos 2000; O'Brien 2000; Tallontire and Blowfield 2000). In the case of the UK food industry, this neo-liberal context is deepened by the laissez-faire national regulatory environment in which the retailers have been permitted both to consolidate and to regulate their supply chains. As a result, ethical trading initiatives developed by UK retailers are shaped by, and in turn shape, a deeply entrenched environment of neo-liberalism and private-interest regulation. This chapter has presented a critique of the corporate strategies adopted by UK supermarket chains to manage the ethical monitoring of their cross-continental food supply chains. Corporate learning in the field of ethical trade has been extremely fast-moving, and I have suggested that three key waves of strategic development have occurred since the mid-1990s. However, as organizational approaches to ethical trade evolve and move through each of these strategic phases – in-house social auditing, 'third-party' auditing and risk assessment – they appear to become increasingly rooted in the commercial drive to defend retailers' brands against the threat of negative publicity, rather than representing the most effective ways of raising labour standards at sites of export production. Coupled with studies of the developmental impacts of labour codes of conduct at sites of production, such research on the changing corporate strategies of key companies in the food system is necessary in uncovering the limitations of private global regulation.

Notes

1   For a more positive perspective on the developmental opportunities afforded by global supply linkages, see Barrett and Browne (1996).
2   The case study material results from a project sponsored by The British Academy on the ways in which retailers are learning to trade ethically. This project involved 27 corporate interviews with the following informants: (i) ethical trading managers

at most of the leading food and clothing retail corporations operating in the UK; (ii) managers of the five largest international audit companies, whose clients include the UK retailers; and (iii) four key ethical consultants involved in advising retailers in matters of ethical trade. The study also involved the author in the participant observation of various spaces of learning for retailers. For the purposes of this chapter, empirical material is drawn from the interviews with management from seven key food retail companies, supported by views from auditors and consultants. The identities of the companies are hidden, in order to respect the wishes of the corporate interviewees.

3 An inquiry into the market power of the UK supermarket chains was conducted by the UK's Competition Commission in 2000, resulting in the recommendation that the retailers should draw up a code of practice to promote the fairer treatment of suppliers. However, there is no obligation on the retailers to extend such a code to cover suppliers overseas (Hughes 2001b).

4 The 1990 Food Safety Act 'rendered retailers legally responsible for the safety of their own-label products' (Marsden and Wrigley 1995, 1996; Doel 1996: 61; Marsden *et al.* 2000).

5 There are also cases where local CSOs and trade unions at sites of export production are involved in more participatory approaches to labour monitoring, for example in regions of South African wine production. However, cases like this are currently the exception rather than the rule (corporate interviews).

6 PricewaterhouseCoopers have now spun off this social auditing function to a firm called Global Social Compliance (O'Rourke 2002).

7 Several interviewees were reluctant to divulge information on their total ethical trading budgets and precise sums of money spent on social audits. Others argued that it is impossible to estimate the precise corporate spends on ethical trade, given that its finance flows through multiple channels within the firm. It is therefore difficult both to produce quantitative data and to draw significant conclusions on the total economic costs (and benefits) of social auditing.

8 Food Retailer F, for example, is engaged in worker education programmes with CSOs at particular sites of export production. This same retailer also works with a more development-orientated external auditor than some of its competitors. Resulting methods of audit and remediation at production sites more strongly favour the long-term development of worker welfare.

References

Adams, R.J. (2002) 'Retail profitability and sweatshops: a global dilemma', *Journal of Retailing and Consumer Services*, 9: 147–53.

Arce, A. and Marsden, T.K. (1993) 'The social construction of international food: a new research agenda', *Economic Geography*, 69: 293–311.

Barrett, H. and Browne, A. (1996) 'Export horticultural production in sub-Saharan Africa', *Geography*, 81: 47–56.

Barrett, H.R., Ilbery, B.W., Browne, A.W. and Binns, T. (1999) 'Globalization and the changing networks of food supply: the importation of fresh horticultural produce from Kenya into the UK', *Transactions of the Institute of British Geographers*, 24: 159–74.

Barrientos, S. (2000) 'Globalisation and ethical trade: assessing the implications for development', *Journal of International Development*, 12: 559–70.

Barrientos, S. (2002) 'Mapping codes through the value chain: from researcher to detective'. In R. Jenkins, R. Pearson and G. Seyfang (eds) *Corporate Responsibility and Labour Rights: Codes of Conduct in the Global Economy*. London and Sterling (VA): Earthscan.

Barrientos, S., McClenaghan, S. and Orton, L. (1999) 'Ethical trade and South African deciduous fruit exports: addressing gender sensitivity', *The European Journal of Development Research*, 12: 140–58.

Beck, U. (1992) *Risk Society*. London: Sage.

Blowfield, M. (1999) 'Ethical trade: a review of developments and issues', *Third World Quarterly*, 20: 753–70.

Blowfield, M. (2002) 'ETI: a multi–stakeholder approach'. In R. Jenkins, R. Pearson and G. Seyfang (eds) *Corporate Responsibility and Labour Rights: Codes of Conduct in the Global Economy*. London and Sterling (VA): Earthscan.

Bowlby, S.R. and Foord, J. (1995) 'Relational contracting between UK retailers and manufacturers', *International Review of Retail, Distribution and Consumer Research*, 5: 333–61.

Cook, I. (1994) 'New fruits and vanity: symbolic production in the global food economy'. In A. Bonnano, L. Busch, W. Friedland, L. Gouveia and E. Mingione (eds) *From Columbus to ConAgra: The Globalization of Agriculture and Food*. Lawrence (KS): University Press of Kansas.

Cook, I. (1995) 'Constructing the 'exotic': the case of tropical fruit'. In J. Allen and D. Massey (eds) *Geographical Worlds*. Oxford: Oxford University Press.

Dannreuther, C. and Lekhi, R. (2000) 'Globalization and the political economy of risk', *Review of International Political Economy*, 7: 574–94.

Doel, C. (1996) 'Market development and organisational change: the case of the food industry'. In N. Wrigley and M. Lowe (eds) *Retailing, Consumption and Capital: Towards the New Retail Geography*. Harlow: Longman.

Ethical Trading Initiative (ETI) (2000) *Getting to Work on Ethical Trading: Annual Report*. London: Ethical Trading Initiative.

Ethical Trading Initiative (ETI) (2002) *Raising the Stakes: Annual Report 2001/2002*. London: Ethical Trading Initiative.

Ethical Trading Initiative (ETI) (2003) 'Ethical trading initiative'. Online: www.ethicaltrade.org/pub/members/list/main/index.shtml (accessed 5 December 2003).

Foord, J., Bowlby, S. and Tillsley, C. (1992) 'Changing relations in the retail-supply chain: geographical and employment implications', *International Journal of Retail and Distribution Management*, 20: 23–30.

Foord, J., Bowlby, S. and Tillsley, C. (1996) 'The changing place of retailer-supplier relations in British retailing'. In N. Wrigley and M. Lowe (eds) *Retailing, Consumption and Capital: Towards the New Retail Geography*. Harlow: Longman.

Freidberg, S. (2003) 'Cleaning up down South: supermarkets, ethical trade and African horticulture', *Social and Cultural Geography*, 4: 27–43.

Freidberg, S. (2004) 'The ethical complex of corporate food power', *Environment and Planning D: Society and Space*, 22: 513–31.

Friedmann, H. (1993) 'The political economy of food', *New Left Review*, 197: 29–57.

Goodman, D. (2002) 'Rethinking food production-consumption: integrative perspectives', *Sociologia Ruralis*, 42: 271–7.

Goodman, D. and Watts, M.J. (eds) (1997) *Globalising Food: Agrarian Questions and Global Restructuring*. London and New York: Routledge.

Grant, R. (1987) 'Manufacturer-retailer relations: the shifting balance of power'. In G. Johnson (ed.) *Business Strategy and Retailing*. Chichester: Wiley.

Hale, A. (2000) 'What hope for 'ethical' trade in the globalised garment industry?', *Antipode*, 32: 349–56.

Hale, A. and Shaw, L.M. (2001) 'Women workers and the promise of ethical trade in the globalised garment industry: a serious beginning?', *Antipode*, 33: 510–30.

Hartwick, E. (1998) 'Geographies of consumption: a commodity-chain approach', *Environment and Planning D: Society and Space*, 16: 423–37.

Hughes, A. (1996) 'Retail restructuring and the strategic significance of food retailers' own-labels: a UK-USA comparison', *Environment and Planning A*, 28: 2201–26.

Hughes, A. (2000) 'Retailers, knowledges and changing commodity networks: the case of the cut flower trade', *Geoforum*, 31: 175–90.

Hughes, A. (2001a) 'Global commodity networks, ethical trade and governmentality: organizing business responsibility in the Kenyan cut flower industry', *Transactions of the Institute of British Geographers*, 26: 390–406.

Hughes, A. (2001b) 'Multi-stakeholder approaches to ethical trade: towards a reorganisation of UK retailers' global supply chains', *Journal of Economic Geography*, 1: 421–37.

Institute of Grocery Distribution (2003) *Grocery Retailing – The Market Report*. Watford: IGD.

Jenkins, R. (2002) 'The political economy of codes of conduct'. In R. Jenkins, R. Pearson and G. Seyfang (eds) *Corporate Responsibility and Labour Rights: Codes of Conduct in the Global Economy*. London and Sterling (VA): Earthscan.

Johns, R. and Vural, L. (2000) 'Class, geography, and the consumerist turn: UNITE and the Stop Sweatshops campaign', *Environment and Planning A*, 32: 1193–213.

McClintock, B. (1999) 'The multinational corporation and social justice: experiments in supranational governance', *Review of Social Economy*, 57: 507–22.

Marsden, T. and Wrigley, N. (1995) 'Regulation, retailing and consumption', *Environment and Planning A* 27: 1899–912.

Marsden, T. and Wrigley, N. (1996) 'Retailing, the food system and the regulatory state'. In N. Wrigley and M. Lowe (eds) *Retailing, Consumption and Capital: Towards the New Retail Geography*. Harlow: Longman.

Marsden, T., Flynn, A. and Harrison, M. (2000) *Consuming Interests: The Social Provision of Foods*, London: UCL Press.

Miller, D. (1998) 'Conclusion: a theory of virtualism'. In J.G. Carrier and D. Miller (eds) *Virtualism: a New Political Economy*. Oxford and New York: Berg.

Miller, D. (2000) 'Virtualism – the culture of political economy'. In I. Cook, D. Crouch, S. Naylor and J.R. Ryan (eds) *Cultural Turns/Geographical Turns: Perspectives on Cultural Geography*. Harlow: Prentice Hall.

O'Brien, R. (2000) 'Workers and world order: the tentative transformation of the international union movement', *Review of International Studies*, 26: 533–56.

O'Rourke, D. (2002) 'Monitoring the monitors: a critique of third-party labour monitoring'. In R. Jenkins, R. Pearson and G. Seyfang (eds) *Corporate Responsibility and Labour Rights: Codes of Conduct in the Global Economy*. London and Sterling (VA): Earthscan.

Pentland, B. (2000) 'Will auditors take over the world? Program, technique and the verification of everything', *Accounting, Organizations and Society*, 25: 307–12.

Power, M. (1997) *The Audit Society: Rituals of Verification*. Oxford and New York: Oxford University Press.

Raynolds, L. (2000) 'Re-embedding global agriculture: the international organic and fair trade movements', *Agriculture and Human Values*, 17: 297–309.

Raynolds, L. (2002) 'Consumer/producer links in fair trade coffee networks', *Sociologia Ruralis*, 42: 404–24.

Raynolds, L., Myhre, D., Figueroa, V., Buttel, F. and McMichael, P. (1993) 'The new internationalization of agriculture', *World Development*, 21: 1101–21.

Renard, M.-C. (2003) 'Fair trade: quality, market and conventions', *Journal of Rural Studies*, 19: 87–96.

Roberts, L. (2002) 'Beyond codes: lessons from the Pentland experience'. In R. Jenkins, R. Pearson and G. Seyfang (eds) *Corporate Responsibility and Labour Rights: Codes of Conduct in the Global Economy*. London and Sterling (VA): Earthscan.

Roberts, S. (2003) 'Supply chain specific? Understanding the patchy success of ethical sourcing initiatives', *Journal of Business Ethics*, 44: 159–70.

Strathern, M. (1997) '"Improving ratings": audit in the British university system', *European Review*, 5: 305–21.

Strathern, M. (2000) 'Introduction: new accountabilities'. In M. Strathern (ed.) *Audit Cultures*. London: Routledge.

Tallontire, A. and Blowfield, M.E. (2000) 'Will the WTO prevent the growth of ethical trade? Implications of potential changes to WTO rules for environmental and social standards in the forest sector', *Journal of International Development*, 12: 571–84.

Tsogas, G. (1999) 'Labour standards in international trade agreements: an assessment of the arguments', *International Journal of Human Resource Management*, 10: 351–75.

Watts, M. (1996) 'Development III: the global agrofood system and late twentieth-century development (or Kautsky *redux*)', *Progress in Human Geography*, 20: 230–45.

Watts, M. and Goodman, D. (1997) 'Agrarian questions: global appetite, local metabolism: nature, culture, and industry in *fin-de-siecle* agro-food systems'. In D. Goodman and M.J. Watts (eds) *Globalising Food: Agrarian Questions and Global Restructuring*. London and New York: Routledge.

Whatmore, S. and Thorne, L. (1997) 'Nourishing networks: alternative geographies of food'. In D. Goodman and M.J. Watts (eds) *Globalising Food: Agrarian Questions and Global Restructuring*. London and New York: Routledge.

Wrigley, N. (1987) 'The concentration of capital in UK grocery retailing', *Environment and Planning A*, 19: 1283–8.

Wrigley, N. (1991) 'Is the 'golden age' of British grocery retailing at a watershed?', *Environment and Planning A*, 23: 1537–44.

Wrigley, N. (1992) 'Antitrust regulation and the restructuring of grocery retailing in Britain and the USA', *Environment and Planning A*, 24: 727–49.

Zadek, S. (1998) 'Balancing performance, ethics and accountability', *Journal of Business Ethics*, 17: 1421–41.

11 The penetration of lead firms in regional agri-food chains
Evidence from the Argentinian fresh fruit and vegetable sector

Mónica Bendini and Norma Steimbreger

Introduction

Argentina's agricultural export base has been historically dominated by livestock and grains. However, in the last two decades it has become larger and more diverse, with the expansion of fruit and vegetables being an important component of these transformations. From 1990 to 1998, the export value of Argentina's six largest fruit and vegetable sectors (apples, pears, oranges, lemons, grapes and dry beans) increased from US$305 million to US$671 million (FAO 2004). In this regard, the Argentinian experience is exemplary of that faced by many countries in the South; fresh fruit and vegetable exports to more affluent Northern countries have been an important and rapidly growing area of the international agri-food system over the past two decades.

Close inspection of the Argentinian case reveals the national-scale dynamics that underpin these processes. Specifically, it emphasizes the role of the 'lead firm' in reorganizing domestic production and forging connections with export markets. Lead firms are companies that seek out and develop new agri-food production sites. For the issues discussed in this book, their importance relates to the way that they extend the geographical and production frontier in agri-food globalization. Analysing agri-food restructuring through the activities of lead firms, therefore, provides a methodological strategy that gives focus to the individual dynamics of global agri-food restructuring and, in particular, to the ways local agricultures are restructured and inserted into international markets (Steimbreger 2001; Steimbreger *et al.* 2002).

In the present case, the expansion of Argentinian fresh fruit and vegetables exports is linked intimately to the expansion of a dominant national firm (Expofrut), which was taken over by a large European company (Bocchi). This chapter documents the corporate strategies used to build this business, and how these have changed over time. From this analysis, it is apparent that Expofrut/Bocchi's fresh fruit export activities have become increasingly sophisticated, with the result that local production has been reorganized to satisfy more exacting requirements from affluent markets. Coincidently, power and control of production arrangements has been shifted increasingly to the company's head office in Europe.[1]

Agri-food globalization and lead firms in Argentinian agriculture

In similar fashion to other Latin American countries, Argentina is experiencing accelerated changes which are characterized by the intensified domination of multinational capital in agriculture. This is impacting on rural labour processes, resulting in more casualized and flexible forms, and is challenging the viability of family farming. There is a deepening of subordinated integration of producers to agri-food chains, where external controls and decisions come from transnational corporations. Taken together, these processes are reconfiguring and redefining the territorial organization of social actors at local level (Bendini and Tsakoumagkos 1999: 1).

In the case of Argentina's fresh fruit and vegetable sector, these processes have been played out in specific geographical contexts. Argentina's export fruit and vegetable sectors are located mainly in areas outside the Pampas. Argentina contributes approximately 4 per cent of global pear and apple production, and pears and apples account for approximately 50 per cent of Argentina's fruit and vegetable exports. More than 85 per cent of pears and 80 per cent of apples are produced in the valleys of the Negro river (Figure 11.1). Citrus products represent a further 40 per cent of fruit exports, of which lemons are the major commodity. Lemons are grown mainly in the province of Tucumán. In terms of world trade, the evolution of the Argentinian fruit exports has been shaped by: international price and production stability in the apple and pear market; increased pear consumption in Europe; the rise in export demand for new bi-coloured varieties of apples; sustained increase of citrus export demand; the opening of the US lemon market and continued growth of some Asian markets; growth in international demand for a wider range of fruits including avocados, mangoes, cherries, berries, figs, nuts and limes; and new export markets for fruit that is organic or produced through integrated systems that make low use of agri-chemicals (*Informe Frutihortícola* 2000).

Capturing these growth opportunities has depended on technical and organizational changes introduced into agricultural chains that modify the appropriation and accumulation of capital. Of particular relevance to this chapter, the increased export of fresh fruits and vegetables has implied changes to a number of post-harvest activities, such as quality control, sorting, conditioning, cooling and packing. These activities increasingly have taken on the character of industrial processes.

These changes have had major implications for the configuration of production systems. Export companies have sought to develop multiple and diverse strategies to enter international commerce. In this period of experimentation, the concept of 'flexibility' has been important. This has restructured traditional production systems and caused increasing levels of concentration and differentiation across sectors and regions. A central element in these processes has been the restructuring of labour processes, through the

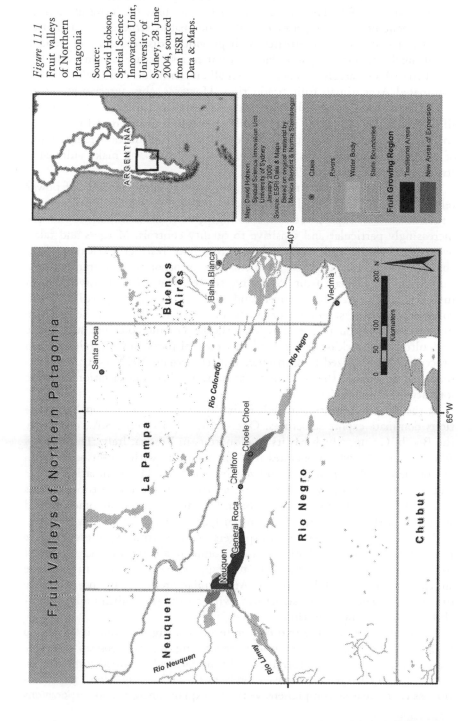

Figure 11.1
Fruit valleys
of Northern
Patagonia

Source:
David Hobson,
Spatial Science
Innovation Unit,
University of
Sydney, 28 June
2004, sourced
from ESRI
Data & Maps.

externalization of services and the subcontracting of workers (Bendini and Tsakoumagkos 1999). These processes redefine not only the forms of internal management of the enterprises but also the spatial structure of activities and the articulation of enterprises with providers, distributors and traders. Additionally, export companies have sought to restructure farming systems. Until the 1990s, largely independent, small and middle-sized family farmers dominated Argentinian fruit production. However, this agrarian structure has been modified during the past decades as export companies have diversified modalities of farming systems within transnationalized and integrated chains. The production contract has been a major tool in the strengthening of vertical integration. This has served to subordinate and make vulnerable small producers (Bendini 1999).

The integration of the Argentinian fruit system expresses a set of power relationships between local production and export markets. In general, Argentinian fruit exports are sold into highly concentrated markets that are increasingly particular and sensitive to quality controls. Mergers and takeovers in wholesale and supermarket sectors have restructured export fruit chains, and have encouraged an increasing incorporation of transnational capital in Argentina's fruit exporting sector. There is extensive foreign investment, via mergers, takeovers and alliances, in the intermediation and commercialization stages of products destined for northern hemisphere markets. These economic groups form the hegemonic core that supply the agriculture and packing sectors and that control ports and sea freight. Such processes, characterized by Constance and Heffernan (1994) as 'true fusion-mania', have created the conditions for rapid market concentration, and the formation of oligopolies.

In the case of Argentinian fresh fruit exports, two international corporations dominate export structures. One of these is Expofrut S.A., owned by the Bocchi Group, which has its headquarters in Verona, Italy. The other is San Miguel S.A., which operates in a joint venture with the Fisher Group, one of the world's largest fresh and vegetable traders. The domination of this sector by these two firms has led to a situation where the fortunes of this increasingly important industry are bound within the corporate practices and strategies of these two firms. Until recently, fruit exporters were well differentiated; they were either citrus exporters or pome-fruit exporters (*Hispano Fruit* 2000). Nowadays, both firms are commercializing a broad variety of fresh products from temperate climates (apples, pears, stone fruit) and the subtropics (citrus fruit). In this way, Argentinian producers are inserted more directly into world market structures, with all the possibilities and vulnerabilities that follow from this.

The emergence of these two corporate groups suggests that if we are to understand this episode of agri-food restructuring, we must possess an appreciation of the structures and strategies of transnational agri-food corporations. Yet although there is an extensive body of research on agri-food globalization, as yet there is no comprehensive theory explaining agri-food corporations

(Pritchard 2000). What is apparent, nevertheless, is that agri-food corporations display considerable prowess for devising new and innovative strategies with which to generate profits. This includes strategies of modernization, vertical coordination, territorial expansion, and alliances and fusions. With regard to fresh fruit and vegetables, this diversity of strategies has been documented by researchers emanating from both the structural traditions of political economy (Friedland 1994, 2001) and post-structural actor network perspectives (Marsden 1999). Evidently, agri-food corporations need to respond nimbly to changing market conditions and, as such, theoretical models attempting to explain these entities must give central understanding to the continuities and ruptures in corporations' organizational forms, as they seek to respond to economic, social and political dynamics (Barbero 1996; Radonich *et al.* 2000). With this in mind, attention now turns to Expofrut/Bocchi, as an example of a lead firm in this agricultural sector.

Trajectory of a lead firm

A local entrepreneur founded Expofrut in 1971. At that time, he was involved in the agricultural business through the sale of tractors and machinery. However, after identifying the potential to supply Europe with off-season fruit, he formed an alliance with a German investor and began to supply this market through a consortium of local fruit producers, packing houses and storage-house owners. From beginnings in the apple sector, Expofrut quickly diversified into other products, including pears, grapes, citrus, onions and garlic.

In the 1970s the firm grew rapidly and, to secure supply, invested in its own farm (*Eurofruit* 1991a, 1991b). By 1981, it had become one of the 'big players' in the fruit business, being ranked third among Argentinian pear and apple exporters (*La Nación* 1999). The purchase of lands for production allowed Expofrut to expand its territorial influence, strengthening its negotiation and price-forming power in the regional market. With its own production lands, Expofrut became less dependent on contract growers.

Additionally, it developed innovative production and commercialization strategies. First, it supplied supermarkets directly, at a time when competitors used the traditional circuits of public auctions, in key sites such as Rotterdam. Second, Expofrut flexibly adapted to changing consumer requirements. It was able to shift its product-mix rapidly, when its competitors in the fruit business were mainly family firms who first chose what to grow and, then tried to place their products in the market.

The 1980s, however, provided a major challenge to these strategies. In 1987, one of its most important customers, the Rewe supermarket chain of Germany, decided to purchase its Argentinian fruit using the Italian trading firm, Bocchi, the largest fruit and vegetable distributor and trader in Europe. Facing these changed conditions, Expofrut agreed to merge its business into the Bocchi group. Bocchi purchased 47 per cent of Expofrut's shares, and

Expofrut sold its exports through Bocchi. This association linked Bocchi's detailed knowledge of European quality and variety requirements to Expofrut's production practices, and gave Expofrut wider market penetration into Europe, leading to increased sales (*Eurofruit* 1991b).

Bocchi itself provides an interesting example of an Italian trading firm. In its structure, characteristics and strategies, it represents a fairly typical example of the middle-sized entrepreneurial firm, usually organized through family networks, that play an important role in Italian commerce (Pritchard and Burch 2003). The company was established in 1966 as an export-import agent coordinating production and trade in the European fruit sector. In 2002 it employed approximately 3,500 workers and sold 1.1 million tonnes of fruit and vegetables, of which 231,000 tonnes came from Expofrut (Expofrut 2002). It has more than 24,282 hectares under production, and owns extensive logistical facilities in South America and Europe. Furthermore, these are inserted within a global operation network that includes subsidiaries distributed in strategic sites in South and North America, Europe, North Africa and Asia. The group has a direct relationship with the large European supermarket chains, some of which are exclusively supplied. Among the Bocchi Group's clients are: Rewe, Spar, Metro, Tegut, Billa, Intermarché, Promodes, Kesko, Ica, Sainsbury's Mercadona, Continente, Prika, Consum and Tuko.

Expofrut's strategic connection with Bocchi enabled the firm to expand considerably. From 1990 onwards, there were substantial changes in the regional organization of agriculture, as Expofrut began its productive expansion towards new areas through land purchases. Additionally, it encouraged greater capitalization of agriculture, implementing new technologies that facilitated Argentinian fruit production to meet European market requirements. Vertical integration and the diversification of production were intensified. Its goals were oriented towards achieving a constant supply of grapes to Europe following a strict schedule according to geographic areas and dates. At that time, Expofrut owned 1,100 hectares under production, mostly in new areas of expansion – mainly with grapes and pears under modern irrigation techniques – and also had citrus production in San Pedro, in the province of Buenos Aires (*Eurofruit* 1991a, 1991b). It then acquired approximately 2,000 hectares of production in northern Patagonia and, in 1992, purchased a 15,000-hectare property in a new area in the province of Río Negro. The expansion to these areas was linked with comparative advantages, such as land quality, location and extension. Apart from the possibility of having big areas with agricultural suitability, the region has adequate urban infrastructure and services, and the proximity to a seaport. All these conditions sealed the circle of advantages for capital investments in this agricultural region (*Diario Río Negro* 12 December 1995). Furthermore, in this way, Expofrut strengthened its presence in the primary segment of the fruit chain. In many regions it had 'quasi-monopoly' powers, being the sole or dominant producer. The corporation developed additional competitive advantages through the incorporation of specialized technology in the whole circuit.

In 1993, the Bocchi Group took majority control of the company by purchasing 92 per cent of the shares. Evidently, Expofrut's capacity to provide counter-seasonal fruit and vegetables for the European market made it an important corporate asset for the Italian group. By 2001, 75.2 per cent of Expofrut's production was exported to countries outside South America (Expofrut 2002). At this time, Expofrut's exports were US$180 million. Its exports to Europe were marketed through Bocchi proprietary brands and supermarket private labels. Fresh pears, apples and citrus that did not achieve export quality standards were sold through discount channels in the domestic market (Expomarket S.A., an Expofrut subsidiary, was established in order to supply the most important Argentinian supermarkets).

Growth and expansion of Expofrut took the company into new down-stream activities. In 1994 the firm purchased primary processing activities for the production of pastes and canned products, and built a fruit packing house. This investment was done within a context in which the rural local authorities were promoting the settlement of industrial enterprises through special offerings, such as tax exemptions (*Diario Río Negro* 12 December 1995). Not only was this the largest fruit processing complex in the country but it was also highly innovative, integrating processing, packing, controlled-atmosphere, cold-storage chambers and quick-cooling tunnels. To further consolidate its market position, in 1997 Expofrut acquired 36 per cent of Terminal de Servicios Portuarios Patagonia Norte S.A., which won a 30-year licence to run the main container port in the province of Río Negro (*Diario Río Negro* 5 August 1997). Also during that year, a new alliance with Citrí-cola Salerno facilitated Bocchi to complete its citrus product variety in the southern hemisphere.

Then in 1999, changes took place in Expofrut's internal management and organization that could be considered as being the most important in its history. The Bocchi Group purchased the remaining 8 per cent of Expofrut's shares so that the company became a wholly owned subsidiary. Following this acquisition, Bocchi restructured local management, leading to a process in which general management, planning, investment and production control were concentrated in home (Italian) locations. The implications of these changes on both the company and Argentinian horticulture are now explored.

Expofrut and the restructuring of Argentinian fruit and vegetable production

By the twenty-first century, Expofrut/Bocchi accounted for approximately 28 per cent of Argentina's total commercial fruit and vegetable production, and 40 per cent of fresh fruit and vegetable exports. Fruit was produced and sold according to well-organized and highly technical systems, based on the establishment of direct sales with supermarket chains in order to eliminate intermediaries and public auctions. This allowed Expofrut to further increase its commercial activities and to plan the development of new varieties.

In the years 2000–1, Expofrut acquired from a bankrupt firm 830 hectares of stone and pome fruit in full production. This investment represented a major territorial acquisition in a region of Argentina in which the firm had not previously operated (the middle valley of the Neuquén river, in the province of Neuquén). For the purposes of corporate strategy, this increased production capacity enabled the company to establish itself more prominently in the domestic market. Moreover, this acquisition took the company's productive land holdings in Argentina to 18,000 hectares, including 3,000 hectares of production in Patagonia, as well as packing and cooling chambers. Within this production Expofrut was able to own-source 50 per cent of its fruit and vegetables, giving the company considerable control over prices and quality parameters (Expofrut 2002). The company's network of activities across Argentina is summarized in Table 11.1.

The impact of Expofrut/Bocchi, however, extends beyond the operations it directly owns. Approximately half of its exports are grown by approximately 450 small and medium-sized farm enterprises. Through production contracts and the lack of commercial alternatives, these producers have remained subordinated to the lead firm. Taking labour conditions as a case in point, there has been public concern about the economic precariousness of casual and seasonal workers, illustrated by demonstrations in 2000 (*Diario Río Negro* 17 November 2000). Off season, the high levels of unemployment

Table 11.1 Expofrut's activities by region, 2002

Region	Activity
North-west region	Business office; lemons, orange and grapefruit packing
Mesopotamian region	Lemon, grapefruit, orange and clementine packing
Cuyo region	Business and administrative office; cold storage plant for grapes and stone fruit (cherries, plums, peaches and nectarines); apples, pears, plum and garlic production
Buenos Aires	Business and administrative offices
Province of Buenos Aires:	
– Campana Harbor	Departure of the first shipments of the season, exporting garlic, grapes, stone fruit and the whole of citrus production
– Mayor Buratovich	Packing of onions; garlic, stone fruit, blueberries and oranges
Northern Patagonia	Pears, apples, stone fruit, grapes and onions; packing houses and cold storage plants.
San Antonio East Port	Cold storage plant and shipping port for 70% of total amount of exports

Source: Expofrut 2002.

and underemployment have required the state to introduce social programmes. These initiatives, however, are inadequate to address the scale of
economic and social problems in some regions.

In summary, the expansion of Expofrut can be said to have been the
linchpin for a set of related transformations in these Argentinian export fresh
fruit and vegetable sectors. These include:

- the incorporation of local management functions into international
  arenas, including the physical relocation of some activities outside the
  country;
- the vertical integration of a significant share of the national fruit and
  vegetable sector, to satisfy the demands of hypermarkets and supermarkets in the importing countries;
- an intensification of technological change in all sectors of the supply
  chain;
- labour restructuring and new forms of intermediation that have encouraged a reduction of permanent workers and an increase of temporary
  ones;
- extensive subcontracting of non-core activities (vertical disintegration),
  at the same time that logistical coordination is being pursued. (Expofrut
  is the only exporter with an independent programme with scheduled
  shipment departures set in accordance with delivery contracts. The
  Bocchi Group coordinates the arrivals and the distribution. The exports
  arrive in the EU through Amberes port and, in the Mediterranean Sea,
  Sagunto and Vado ports. During the low season, the exported citrus
  arrive in Rotterdam port (Expofrut 2002);
- new configurations of the spatial structure of Argentine fruit and
  vegetable production by way of territorial expansion in new production
  areas;
- the widespread use of contract agriculture with primary suppliers and
  with packing houses; and
- the consolidation of a national product supply network of fresh fruit and
  vegetables.

Conclusion

A focus on the activities and strategies of lead firms provides a useful methodological and theoretical approach for understanding global agri-food
restructuring and the creation of cross-continental food chains. In the present
case, the example of Expofrut/Bocchi reveals a succession of changes in
Argentinian fruit and vegetable production that have led to greater and more
diverse forms of penetration of international capital in the local economy.
Expofrut/Bocchi is a hegemonic actor that has materially changed patterns
of capital accumulation in the Argentinian fruit system. This has involved
the transnationalization of supply chains, greater concentration of economic

resources and the social subordination of agricultural workers and contract farmers.

The transformative effects of Expofrut/Bocchi in the Argentinian fresh fruit and vegetable sector represent the specific modalities by which transnational capital articulates with local regions (Bonanno *et al.* 1999; Bendini 2002). This narrative is at once both globally relevant – emphasizing important general tendencies in the global agri-food system – and particularistic – in that it highlights specific historical conditions in Argentina's agricultural political economy. Furthermore, it brings into focus the cumulative changes and global–local tensions that accompany agri-food restructuring. With reference to Latin America, this case study documents the changing power structures that are integral to the most recent phase of agrarian restructuring. Expofrut/Bocchi represents a new manifestation of the classic Latin American plantation agro-economy (Murmis and Bendini 2003), in which hegemonic interests shape the social conditions of agriculture in specific local spaces.

Notes

1   The material presented here is derived from a comprehensive study of Expofrut/
    Bocchi's operations in Argentina, including interviews with managers, workers, offi-
    cials and farmers, and the use of secondary sources such as company reports and
    Internet resources. The article uses findings from the research project, M. Bendini
    et al., UNCo – PICT 04 747, *Cambios en la cadena de valor frutícola y reposicionamiento
    de productores, empresas y trabajadores* and from the master's thesis of N. Steimbreger,
    *Trayectoria y reorganización de una empresa frutícola en el marco de la reestructuracíon produc-
    tiva*, Sociology of Latin American Agriculture, College of Law and Social Sciences,
    National University of Comahue, Argentina.

References

Barbero, M. (1996) 'Treinta años de estudios sobre la historia de empresas en la
    Argentina'. In F. Cimin and P. Aceña (eds) *La Empresa en la Historia de España*. Madrid,
    Editorial Civitas.
Bendini, M. (1999) 'Entre macas e peras: globalizacao, competitividade e trabalho'. In
    J.S.B. Cavalcanti (ed.) *Globalizacao, Trabalho e Meio Ambiente: Mucanzas Socioeconomicas
    em Regioes Frutícolas para Exportacao*. Recife (Brazil): Editora Universitária UFPE.
Bendini, M. (2002) 'La configuración de una región agrícola dinámica: Actores sociales
    en la negociación local'. In J.M. Sumpsi and E. Pérez (eds) *Políticas, Instrumentos y
    Experiencias de Desarrollo Rural en América Latina y Europa*. Madrid, AECI-FODEPAL-
    MAPYA.
Bendini, M. and Tsakoumagkos, P. (eds) (1999) *Transformaciones Agroindustriales y
    Laborales en Nuevas y Tradicionales Zonas Frutícolas del Norte de la Patagonia, Cuaderno
    GESA 3 y PIEA 10*. Buenos Aires: Universidad de Buenos Aires.
Bonanno, A., Marsden, T. and Graziano da Silva, J. (1999) 'Globalizacao e localizacao:
    elementos para entender a reestruturacao dos espacos rurais'. In J.S.B. Cavalcanti (ed.)
    *Globalizacao, Trabalho, Meio Ambiente. Mucanzas Socioeconomicas em Regioes Frutícolas para
    Exportacao*. Recife (Brazil): Editora Universitária UFPE.

Constance, D. and Heffernan, W. (1994) 'Las empresas transnacionales y la globalizacion del sistema alimentario'. In A. Bonanno (ed.) *Globalización del Sector Agrícola y Alimentario: Serie Estudios*. Madrid, Ministerio de Agricultura, Pesca y Alimentación–Secretaria General Técnica.

Diario Rio Negro, 12 December 1995 and 17 November 2000, Río Negro, Argentina.

Eurofruit (1991a) 'Argentina production yet to reach full potential, conference told', *Eurofruit*, July: 12–16.

Eurofruit (1991b) 'The Bocchi connection: Argentine exporter Expofrut has seen sales rocket in Europe since it teamed up with Bocchi Food Trade International', *Eurofruit*, July: 20–1.

Expofrut (2002) 'Expofrut'. Online: www.expofrut.com.ar (accessed 1 May 2003).

Food and Agricultural Organization (FAO) (2004) 'FAOSTAT database'. Online: http://faostat.fao.org (accessed 26 May 2004).

Friedland, W.H. (1994) 'The new globalisation: the case of fresh produce'. In A. Bonanno, L. Busch, W.H. Friedland, L. Gouveia and E. Mingione (eds) *From Columbus to ConAgra: The Globalization of Agriculture and Food*. Lawrence (KS), University of Kansas Press: 210–31.

Friedland, W.H. (2001) 'Reprise on commodity systems methodology', *International Journal of Sociology of Agriculture and Food*, 9 (1): 82–103.

Hispano Fruit (2000) 15 April.

Informe Frutihortícola (2000) 15 (176), February.

La Nación (1999) 'La Nación Online'. Online: www.lanacion.com.ar (accessed 3 January 2003).

Marsden, T. (1999) 'Globalizacao e sustentabilidade: criando espacos para alimentos e naturaleza'. In J.S.B. Cavalcanti (ed.) *Globalizacao, Trabalho e Meio Ambiente: Mucanzas Socioeconomicas em Regioes Frutícolas para Exportacao*. Recife (Brazil): Editora Universitária UFPE.

Murmis, M. and Bendini, M. (2003) 'Imágenes del campo latinoamericano en el contexto de la mundialización'. In M. Bendini, J.S.B. Cavalcanti, M. Murmis and P. Tsakoumagkos (eds) *El campo en la sociología actual: una perspectiva latinoamericana*. Buenos Aires: Editorial La Colmena.

Pritchard, B. (2000) 'The tangible and intangible spaces of agro-food capital', unpublished paper presented at *X Congress of the International Rural Sociological Association*, Rio de Janeiro, July (available from the author: School of Geosciences, University of Sydney, 2006 NSW Australia).

Pritchard, B. and Burch, D. (2003) *Agri-food Globalization in Perspective: International Restructuring in the Processing Tomato Industry*. Aldershot: Ashgate.

Radonich, M., Steimbreger, N., Miralles, G. *et al.* (2000) 'En el ritmo de la fruticultura: la empresa como director de orquesta', ANALES de la Sociedad Chilena de Ciencias Geográficas, Santiago de Chile.

Steimbreger, N. (2001) *Trayectoria y Reorganización de una Empresa Frutícola en el Marco de la Reestructuración Productiva*. Unpublished masters' thesis, Universidad Nacional del Comahue, Argentina.

Steimbreger, N., Castañón, M. and Vecchia, M. (2002) *Procesos de Diferenciación Empresarial en la Fruticultura del Norte de la Patagonia*. Unpublished. *Paper III Jornadas de Extensión del Mercosur*. Buenos Aires: AADER.

12 Production, consumption and trade in poultry

Corporate linkages and North–South supply chains

David Burch

Introduction

Typically, the analysis of changes in patterns of production, consumption and trade in agricultural commodities is based on statistics which are aggregated at the national level. While such formulations are useful in identifying trade flows between nation-states, they are less relevant in a period when over one-third of world trade occurs within the boundaries of individual firms (Johnson and Turner 2003: 101). When analysing the global trade in poultry for example, what does it mean when it is stated that 'Japan will source more of its poultry imports from China as opposed to Thailand' (Foreign Agricultural Service (FAS) 2003a), if the same company, operating in both China and Thailand, is the supplier in both instances? And how does an analysis of trade flows at the national level capture the realities of international coordination by supermarkets and fast food companies, which are able to operate in – and source products from – an array of production sites?

These questions suggest that in order to fully understand the dynamics of contemporary change in agri-food supply chains, it is necessary to adopt an alternative framework in which global production, trade and consumption are analysed from the perspective of the corporate entities involved as well as the nation-state in which production facilities may be located. To this end, this chapter examines recent international restructuring in the poultry industry through the activities of two transnational corporations – the Charoen Pokphand Group (CP Group) with its origins in the 'South', and the Grampian Country Food Group (GCFG) which was established in the 'North'. Both of these firms have been involved in significant restructuring that has included the establishment or acquisition of productive facilities outside their home base. Specifically, the chapter asks:

1 What is driving geographical shifts in production? Are companies expanding operations in the 'South' in order to satisfy growing domestic demand in these countries, or to develop low-cost production sites in order to supply 'Northern' markets, or both?

2   What do these developments tell us about the flexibility possessed by
poultry producers in selecting new production sites, and how far is this
process likely to go? How does this current restructuring compare
with the capacity of other companies – in the manufacturing sector in
particular – to operate in a 'footloose' and flexible way?

The global poultry industry

The remarkable development of the poultry system in the post-1945 period
saw the chicken transformed from a luxury product to an item of everyday
consumption in most of the developed world (Dixon 2002). In the US, annual
per capita poultry consumption more than doubled from 1975 to 2000, from
16 kilograms to nearly 37 kilograms. In Europe, Eire currently leads with
per capita consumption of 32.1 kilograms, followed by Portugal (31.9 kilograms) and the UK (28.8 kilograms). As would be expected, consumption
in less developed countries is much lower. In China, per capita consumption
is currently around 4.8 kilograms, in the Philippines it is 7.1 kilograms, and
in Thailand it is 11.9 kilograms (FAS 2002c; 2003a).

This consumption is satisfied by production in both the 'North' and
'South'. In 2003, approximately 28 per cent of world poultry production
took place in the US, followed by China (19 per cent) and Brazil (11 per
cent). Growing consumption in the less developed countries has been met

Table 12.1 Poultry and poultry products: exports from selected countries (million
metric tonnes and per cent of world total)

	1995	1997[a]	1999	2001	2003[p]
US	1.77	2.12	2.08	2.52	2.24
	(45.7%)	(44.8%)	(40.9%)	(45.1%)	(36.8%)
Brazil	0.42	0.65	0.74	1.23	1.9
	(11.0%)	(13.8%)	(14.5%)	(22.0%)	(31.3%)
China	0.29	0.35	0.38	0.49	0.39
	(7.5%)	(7.4%)	(7.4%)	(8.7%)	(6.4%)
Thailand	0.17	0.19	0.29	0.42	0.53
	(4.5%)	(4.1%)	(5.6%)	(7.6%)	(8.7%)
World	3.86	4.72	5.09	5.59	6.08

Source: Foreign Agricultural Service (2004).

Notes:

a   As from 1997, chicken feet/paws are not included in the data. While this may affect a reading
of the comparative data between 1996 and subsequent years, this has little impact on the data
that show a declining share of world exports for the US after 1997.

p = Provisional.

both by increased local production and higher levels of imports, but as Table 12.1 shows, the 'South' has also come to account for a growing share of global exports. At the same time, trade in poultry products as a percentage of total production is relatively small. In 2002, only about 11 per cent of world production entered international trade. There are, of course, variations around these data. Thailand, for example, exported 34 per cent of its poultry production in 2002, while in the same year Japan imported some 38 per cent of its total consumption (FAS 2002b; 2002c).

Poultry in the 'South': the CP Group in Thailand and China

Thailand was one of the first 'Southern' countries to emerge as a major site for fully integrated poultry production and export, and in 2003 it was the world's seventh largest producer and the fourth largest exporter. The company that pioneered the industry in Thailand was the CP Group, which was originally established as a small trading and supply company in Bangkok in 1921 by two Chinese migrant brothers. In 1973, the CP Group established the Bangkok Farm Company, which laid the basis for the subsequent expansion of the Group and its emergence as the largest agro-industrial company in Asia, with operations in livestock, fruit and vegetables, grain and feed products, telecommunications, property development, insurance, motorcycles, carpets, convenience stores and supermarkets, shopping malls and fast food outlets. In 2002, the CP Group had a turnover of US$13 billion, and a workforce of 100,000 (not including many thousands of contract farmers) in over 250 companies in 20 countries. However, poultry production and processing and its associated activities (e.g. animal feed production and breeding facilities) remain the most important areas of the Group's activities, accounting for over 65 per cent by value of the total output in 2002. In addition to its production base in Thailand, the CP Group has established poultry production and processing facilities in Turkey, China, Malaysia, Indonesia and the US, as well as animal feed operations in India, China, Indonesia and Vietnam.[1] The CP Group is the world's fourth largest poultry producer (after the US firms Tyson Foods, Perdue Farms and Goldkist) and is the world's largest producer of animal feed (Burch and Goss 1999; Goss and Burch 2001; Goss 2002; *Reuters News* 25 September 2002).

The CP Group is unique in the global poultry industry because of the scope of its operations, the diversity of its non-agricultural activities and its commitment to the vertical integration of its agri-food operations 'from the seed to the supermarket'. In poultry, the Group has operated a large number of fast food outlets, beginning with the KFC chain in Thailand and China, and including, more recently, the 'Chester's Grill' chain in Thailand and the Thai-themed 'Bua Baan' restaurants in Thailand and China. The CP Group also operates over 2000 7-Eleven convenience stores in Thailand, as well as some 17 Lotus superstores in China (with plans to expand this number to 100 stores by 2006).[2] This degree of vertical integration not only means that the Group is involved in the whole supply chain, from the breeding of chicks to the sale of chicken products in its restaurant outlets and retail stores, but it is also well placed to supply processed poultry products to other outlets within Thailand and overseas. For example, the Group has operated as the main supplier of poultry products to Pizza Hut and McDonald's in Thailand, as well as to KFC in Singapore and the UK (Burch and Goss 1999; *The Guardian* 8 July 2002; *Bangkok Post* 28 April 2003).

The CP Group is also a major supplier to a number of retail outlets in the EU, and the UK in particular, as a result of its association with Tesco, the UK's largest supermarket chain and a major player in the emerging global retail sector. This relationship was established in the aftermath of the Asian economic crisis of 1997, when the CP Group was forced to dispose of some of its assets in order to service its debts. In 1998, the CP Group sold 75 per cent of its holdings in Lotus superstores in Thailand to Tesco, and later reduced its remaining 25 per cent share of the company to only 1 per cent. Nevertheless, the Group established itself as a major supplier of food products for sale in Tesco's UK and European outlets. By 2002 the CP Group was exporting chicken products valued at US$24 million to Tesco UK, and this rose to US$127 million by 2004. According to some reports, this latter figure represented 60 per cent of the value of the CP Group's exports to the EU, and between 30 and 40 per cent of the Group's total export revenue (*Bangkok Post* 30 October 1999, 19 February 2000; *The Nation* 11 May 2002; *Thai News Service* 12 February 2004).

The dynamics of production integration are demonstrated particularly well by the CP Group's operations in China. China is both a major exporter and importer of poultry products, although the pattern of trade is quite distinct. In 2002, 77 per cent by volume of the imports of poultry into China consisted of paws, wingtips and offal, which are preferred by most Chinese consumers. The US accounts for 60 per cent of poultry imports into China, and the market for these cheaper cuts is important for the sale of products for which there is little demand in the US. In contrast, some 92 per cent of poultry exports from China consist of higher-value cuts, including whole and portioned broilers and value-added processed products. The largest market for poultry exports from China is Japan, which accounts for 69 per cent of exports (FAS 2003a; 2003b).[3]

The CP Group played a seminal role in the construction of this export–import complex. It was the first major foreign investor in China following economic liberalization in the 1980s, and by 2002 it was the largest operator in the country with joint-venture processing facilities in Qingdao, Shanghai, Beijing, Qinhuangdao, Heilongjiang and Jilin.[4] In 2001, these facilities produced 600 million of the 2.2 billion chickens sold commercially in China, representing 27 per cent of total production. In addition, the CP

Group has further investments in the Chinese poultry sector via its Taiwanese subsidiary, Charoen Pokphand Enterprise, which invested US$12 million in animal feed facilities and chicken farms at Shenyang in Liaoning province, and Lianyun in Jiangsu province (*Taiwan Economic News* 21 May 2002).

In China, as in Thailand, the CP Group has sought to incorporate its poultry activities within vertically integrated structures. In terms of upstream investments, the company has established 109 feed mills in China, spread across 29 of the country's 31 provinces. These operations produce 8 million tonnes of animal feed per annum, with domestic sales satisfying 9 per cent of China's demand consumption, and export sales comprising 10 per cent of China's feed grain exports. Downstream, the Group has operated the KFC franchise in 13 of China's largest cities, which, in the late 1990s, involved the annual delivery of 75.5 million birds. As noted earlier, the Group also established its Lotus supermarket chain in China (which it retained while selling the Lotus supermarkets in Thailand to Tesco), and the 'Bua Baan' chain of Thai-themed restaurants. The first of these restaurants was opened in 2002 in the Super Brand Mall in Shanghai, the largest mall in Asia, also owned by the CP Group (Goss *et al.* 2000; Goss 2002; *Bangkok Post* 16 May 2002).

In addition to its domestic operations in China, the CP Group is centrally involved in the export of poultry from China, and by 2002 its poultry exports from China exceeded those from its facilities in Thailand. This has given the CP Group a considerable degree of flexibility in sourcing, which has become particularly important in light of major price fluctuations in recent years (Goss 2002; *Reuters News* 25 September 2002). For example, when Japan banned poultry imports from China in May 2003 following an outbreak of avian influenza, export prices to Japan more than doubled, increasing from US$1,300 per tonne in March 2003 to US$3,000 in mid-July. On this occasion, Japanese buyers switched the supply of 200,000 tonnes of poultry from China to Thailand. While the volume of orders lost by the CP Group's China operations is not known, this shift in sourcing meant that the CP Group in Thailand obtained additional orders for 30,000 tonnes of poultry at US$2,900 per tonne. Of course, as a consequence of the supply situation, export prices to other markets also rose, and CP Foods was receiving US$2,400 per tonne on its forward contracts with buyers from the EU, up from US$1,900 in the first quarter of 2003 (*Bangkok Post* 9 August 2003, 3 November 2003).[5]

The experience of the CP Group in China clearly demonstrates that South-to-South foreign direct investment is not just about gaining access to ever lower production costs but involves sophisticated corporate strategies of international production flexibility and the construction of vertically integrated production systems in fast-growing markets. Equally importantly, the CP Group's investments in the retail sector have reduced its dependence on international retail and food service companies, with their intense pressures on price margins (Hughes 1996; Burch and Goss 1999; Vorley 2001; Cox *et al.* 2002; Wilkinson 2002; Dobson 2003).

Poultry in the 'North' and the 'South': GCFG in the UK and Thailand

The GCFG was established in Banff, north-east Scotland, in 1980. From a single plant operation employing about 50 people, GCFG has grown to be Britain's largest private unlisted agri-food company, supplying a range of meat products to the UK retail, wholesale and food service sectors. By 2003, the company was operating in 43 locations (39 in the UK, one in continental Europe and three in Thailand), employed over 21,000 people, and had a turnover of £1.4 billion (approximately US$875 million). The company vertically integrates key upstream poultry activities (laying farms, hatcheries, feed mills and processing facilities), and uses contract producers to 'grow out' day-old chicks. It processes approximately 3.8 million chickens per week, which translate into 5,600 tons of whole or portioned chickens, as well as 470 tons of value-added products (chicken strips, satay sticks, etc.), and 140 tons of cooked chicken every week (GCFG 2004).

For its first two decades the GCFG grew via investments and acquisitions in the domestic UK market, but in December 2001 it expanded internationally when it purchased Golden Foods International, a Thai poultry firm owned previously by Wessanen, a Dutch agri-business company. With this acquisition, the GCFG took control of two processing plants near Bangkok (at Pathumthani and Nakhon Nayok), two hatcheries producing 500,000 day-old chicks per week, a feed mill and a distribution and marketing facility in Germany that handled European sales. The Thai operation, renamed 'Grampian Siam', had a capacity of 270,000 birds per day, employed 4,000 workers as well as large numbers of contract growers, and exported 80 per cent of its output, mainly to Japan and Europe. At the time of the takeover, Grampian Siam was able to produce over 200 different chicken dishes and had a weekly output of some 1.5 million stick-based products such as kebabs. Soon after the takeover, GCFG announced a £25 million (US$15.6 million) investment programme over the period 2002–5, which was intended to increase capacity from 800,000 to 1.2 million birds per week (*Aberdeen Press and Journal* 21 December 2001, 23 May 2002).

This investment occurred at a time when UK processors were being increasingly exposed to competition from cheaper imports from the 'South', mostly Thailand and China. In 2003, imports accounted for 31 per cent of UK consumption of poultry and poultry products (Department of Environment, Food and Rural Affairs 2004). Thailand's share of these imports had grown from 3.3 per cent of the total in 1999 to 11.8 per cent in 2002, as companies such as KFC and Tesco sourced supplies of chicken strips and other value-added products from the CP Group and other processors. Such imports posed a direct threat to GCFG's position in the UK as a supplier of these same products to Tesco and KFC. An important factor in this situation was that the CP Group was able to secure labour at the minimum wage of 165 baht (US$3.75 per day at then-prevailing rates), while UK companies had much higher costs in what was a very labour-intensive process (*Agra Europe* 21 December 2001; *Guardian* 8 July 2002; *Aberdeen Press and Journal* 22 March 2003).[6]

In September 2001, just two months before it acquired the Thai operations, the GCFG announced the closure of the chicken-processing facility at Garstang in Lancashire, which involved the loss of 240 jobs. Then, in May 2002, at about the same time that the GCFG announced its plans for increased investment in Thailand, the company also announced the closure of its plant at Newbridge, Edinburgh, with the loss of 547 jobs. On this occasion, the GCFG stated that the plant could not compete with imports from Brazil and Thailand, and that building anew on the existing site was uneconomic (*Lancashire Evening Post* 21 December 2001; *Business a.m.* 28 May 2002).

As to the future, there seems little doubt that the policy of expansion through the acquisition of existing UK companies will give way to a strategy of concentrating future growth in the 'South'. Despite a 61 per cent growth in company profits between 2000 and 2003, the GCFG was still registering a rate of return on turnover of only 2.7 per cent (*The Herald* 3 April 2003; *Sunday Times* 4 April 2004). The company operated in a highly competitive environment as a supplier to supermarket and fast food chains, which were continually looking to reduce costs by reducing the prices paid to domestic suppliers. This has been manifested in a progressive thinning of margins all the way along the supply chain. By 2004, it was being reported that UK poultry producers were receiving 49.5 pence per kilogram for a commodity which cost them 51.5 pence per kilogram to produce (*Farmer's Weekly* 6 February 2004). While attempting to negotiate an increase of 8 pence per kilogram from reluctant retail chains, the Chairman of GCFG concluded that 'in the face of such intransigence, the only thing Grampian and its peers can do is improve efficiencies'. In looking to such a solution, he went on to suggest that 'we have to see ourselves as a global sourcing entity' which would have to look overseas – to Latin America, Eastern Europe and Australasia – for further growth (*Sunday Herald* 21 March 2004).

Such developments provide important insights into the growing market power of international retailers, as companies such as Tesco, Carrefour and Wal-Mart increasingly take responsibility for organizing and managing agri-food supply chains, and pressure their suppliers for ever cheaper product lines (Hughes 1996; Burch and Goss 1999; Vorley 2001; Dobson 2003). The pressures brought to bear on the UK poultry processors are experienced by all agri-food manufacturing companies to a greater or lesser degree. The important element in determining the capacity of any agri-food company to exert influence in the supply chain depends upon the power relationships existing between participants in the supply chain, and between manufacturers and retailers in particular. These power relationships, in turn, depend upon a number of variables associated with the scale of operations of participants, the extent of retail concentration and the existence of alternative marketing opportunities (Cox *et al.* 2002). From this perspective, the GCFG would appear to operate in a very difficult environment with little control over the supply chain and minimal influence in setting the terms of the relationships within this chain. For example, unlike the CP Group, the GCFG neither operates its own retail outlets nor exports poultry products. It is mainly a supplier to supermarket chains and fast food restaurants in the UK market, with few options in terms of other marketing outlets. Moreover, the UK market relies heavily on imports, which means that the GCFG and other processors are always exposed to the threat of cheap imports from offshore competitors.

Conclusion

There seems little doubt that there is occurring a significant shift in poultry production from the 'North' to the 'South'. While overall global production is increasing, the share of world exports accounted for by the two leading 'Northern' exporting countries (the US and France) declined from 54.9 per cent in 1995 to 42.4 per cent in 2002. Over the same period, the leading 'Southern' producers (Brazil, China and Thailand) increased their share, from 23.0 per cent in 1995 to 46 per cent in 2003 (Table 12.1). This shift has been financed by both local and overseas capital, and clearly there is a move by a number of the world's leading poultry processing companies to the 'South', in order to service expanding local markets and to establish export production platforms. This shift has usually involved the acquisition by European and US companies of an existing operation, i.e. a 'Southern' processing company (e.g. GCFG and Golden Foods in Thailand; the US-based Tyson Foods and Nochistongo in Mexico) or the establishment of a joint venture arrangement with local companies that are already engaged in the poultry sector (e.g. the CP Group and Beijing Dafa Livestock Corporation in China; and US companies Cargill and Sadia in Brazil, Tyson Foods and Zucheng Da Long in China and Perdue Farms and Dah Chong Hong in China). In terms of future developments, there appears to be little reason why this shift in the location of production should not continue. As noted earlier, the evidence from the UK is that the GCFG is increasingly looking overseas for its sources of raw materials or for prepared chicken products, while US companies continue to seek opportunities in leading 'Southern' producers such as Brazil, Thailand and China. Undoubtedly, even lower-cost sites of production, such as the Philippines and India, are also likely to emerge as global suppliers in the future (FAS 2002a).

However, the reasons for this shifting pattern of exports are complex, and involve more than just the issue of lower production costs. There is a series of interrelated causal factors, such as the environmental implications of poultry production, which cannot be discussed here for reasons of space.[7] But the most important of these is the tightening of margins and profit conditions in 'Northern' markets. As illustrated in the case of the GCFG, poultry processors in the UK face intense competition from offshore competitors, and are in relatively weak positions in supply chains where market power rests increasingly with supermarkets and fast food companies.

Such speculations direct attention to the second question at the start of this chapter, concerning the geographical mobility and flexibility of large global agri-food companies. In this context, one of the central issues relates to the suggestion that because agri-food production systems rely upon nature, and are dependent upon the specificities of place and climate, they are restricted in their geographical flexibility and in their ability to operate in the same way as 'footloose' industrial companies. While most commentators have discussed this in the context of land-based (i.e. crop) production systems, this issue has also been introduced (albeit in a modified form) by Boyd and Watts (1997) who, in their analysis of the emergence of the modern poultry industry in the southern states of the US, ask how 'the irreducible biological character of the chicken shaped and constrained the organisation of production in the industry?' However, several years later, as US companies are closing poultry processing facilities in the southern United States at the same time as they are expanding operations in China, Brazil and Mexico, this has ceased to be the key question.[8] Instead, we should be asking how and why, and with what degree of flexibility, are major US processing companies able to relocate production capacities from previously favourable locations in the southern United States to new sites in the less developed countries?

It might then be argued that the 'organic' attributes of systems of food provisioning do not apply uniformly across all agri-foods sectors, or in all circumstances, and this is particularly so in the case of the modern poultry industry. This commodity is produced within a closed system, and relies upon a wide range of standardized inputs which can be modified and controlled in order to maximize efficiencies and reduce uncertainty in the production of a predictable and standardized product. Inputs such as day-old chicks are made uniform by a process of genetic manipulation, and raised in a controlled climate on an optimal and regulated diet. The end result is a fully grown chicken, produced in some 40 days, monitored and manipulated in order to ensure the production of a high quality and predictable commodity designed to conform to certain standards.

While nature can never be entirely eliminated from such a system, nevertheless the specificity and unpredictability of 'nature' can be so significantly reduced that, to all intents and purposes, what we see in the modern system of poultry production is an industrial process, which is as flexible and mobile as that in any manufacturing industry. Such a closed system can be established almost anywhere in the world. Indeed, such facilities have been constructed almost everywhere in the world, with little evidence of biological limitations.[9] In fact, it appears to be the case that the constraints to mobility which do exist are more likely to be socio-cultural and political than biological. For example, when the processing companies discussed in this chapter first initiated overseas operations, they did so either by taking over an existing company or by entering into a joint venture with an existing company. Few 'Northern' companies have ever become involved in the 'South' without some local involvement. This acknowledges that the cost of establishing a new poultry processing enterprise overseas goes beyond the provision of capital and physical infrastructure, and the transfer of technology; it also involves knowledge of local conditions concerning markets and marketing capacities, labour practice and policy, government support, financial institutions, social values and more. These are the 'specificities of location', rather than the biological basis of poultry production, which are the issues that have to be addressed as the poultry industry continues to relocate to the 'South' in response to a continuing demand for cheap, quality products all over the world.

Notes

1   The US operation was sold in 2003.
2   The Group sold its interest in KFC restaurants in Thailand in 2000 in order to concentrate on the development of its Chester's Grill chain of restaurants (*Bangkok Post* 2001).
3   The situation in China is complicated by the fact that data on production, consumption, imports and exports for Hong Kong continue to be treated separately from those of mainland China. While Hong Kong has a substantial poultry industry, most of its requirements are met with imports. These totalled 837,000 metric tonnes in 2001, with 66 per cent of this supplied by the US. However, 82 per cent of the US supply, mostly in the form of chicken paws, wings and similar cuts, were re-exported to China. Similarly, there is a large but unspecified volume of trade in US poultry that is imported into China and re-processed before being exported to Japan (FAS 2003a).
4   The CP Group still holds the very first Investment Permits (Number 00001) for both Shenzhen and Shantou special economic zones, issued in 1979, and it remains one of the largest single foreign investors in the country, with over US$5billion invested in some 130 joint ventures. US$4 billion of this is invested in the feed and poultry sector (Ngui 2001; Handley 2003). Note that in 2001, the CP Group began to dispose of its shares in the Shanghai Dajiang plant, and by March 2003 it maintained only 2.8 per cent of the company's equity (*China Daily* 14 March 2003).
5   Similar flexibility in sourcing was demonstrated in early 2004, when a new strain of avian flu, H5N1, emerged to infect the chicken populations of a number of Asian countries, including Thailand, Indonesia, China and Vietnam, which resulted in the banning of imports of fresh and frozen chickens from infected countries by Japan, the EU and others. As a consequence, a number of fast food restaurants in Asia, including McDonald's in Japan and the KFC franchisees in Japan, Hong Kong and Indonesia, ceased buying poultry products from the usual supplier, the CP Group in Thailand, and instead placed orders with the CP Group's Taiwanese subsidiary (*Reuters News* 5 February 2004).
6   The problem of labour supply was one that impacted upon processing companies in most of the major producers in the 'North'. Companies in the UK did seek to reduce

labour costs there, in particular by employing immigrant workers on a casual basis (*The Guardian* 8 July 2002).

7   For example, the question of the environmental impacts of poultry production and processing is a major issue in the 'North', where the industry is increasingly being held accountable for the pollution of land and water resources, and the blighting of urban communities as a consequence of odours and other outputs from farms and factories. These concerns have resulted in a series of well-publicized court actions in the US in recent years, which have cost the industry dearly in terms of compensation and environmental repair. In 2003, in one of the largest actions, Tyson Foods pleaded guilty to twenty violation of federal clean water legislation at its Sedalia, Missouri, poultry plant, and was fined US$7.5 million. More recently, the company was accused by the Sierra Club of failing to report ammonia emissions which were released from four of its poultry facilities in Kentucky. The company was found to be liable in this case. *The Arkansas Democrat Gazette*, 23 October 2003, 13 December 2003; US Environmental Protection Agency, 3 July 2003; *Hazardous Waste Litigation Reporter*, 21 November 2003.

8   For example, between December 2002 and February 2004, Tyson Foods closed 5 of its 54 poultry processing plants in the southern US, and scaled back production at a sixth plant, resulting in a reduction in output of at least 2.36 million birds per week (about 4.5 per cent of the company's production). These plant closures coincided with Tyson's expansion into China and elsewhere, and resulted in the loss of some 3,300 jobs in the company's US processing plants, and the dropping of hundreds of contract growers. See *Memphis Business Journal* 6 December 2002; *Business First of Louisville* 9 December 2002; *Business Journal of Jacksonville* 24 December 2002; *Washington Business Journal* 21 April 2003; *Arkansas Democrat Gazette* 25 October 2003; *Associated Press Newswires* 10 January 2004; Tyson Foods 2004; Stevens *et al.* 2003.

9   It has been suggested that to some extent, the outbreaks of a variety of strain of avian influenza and other diseases that affected large numbers of chickens in numerous production sites from the late 1990s can be attributed to the artificial nature of modern intensive production systems and the fact that large numbers of chickens sharing the same space are bound to be susceptible to disease. Such outbreaks may be more likely to occur as the industry shifts location to the less developed countries, where local chicken growers operate to lower standards and are more likely to infect domestic flocks, but the converse argument also suggests that since avian influenza is contracted from chickens that are in close proximity to people, the more enclosed and isolated a chicken processing facility is from the backyard operations of small peasant producers, the less likely it is that the disease will spread to the modern facilities.

References

Aberdeen Press and Journal (Aberdeen), 21 December 2001; 23 May 2002, 22 March 2003.

Agra Europe (Tunbridge Wells), 21 December 2001.

Arkansas Democrat Gazette (Little Rock), 23 October 2003; 13 December 2003.

Associated Press Newswires (New York), 10 January 2004.

Bangkok Post (Bangkok), 30 October 1999; 19 February 2000; 22 August 2001; 16 May 2002; 28 April 2003; 9 August 2003; 3 November 2003.

Boyd, W. and Watts, M. (1997) 'Agro-industrial just-in-time: the chicken industry and postwar capitalism'. In D. Goodman and M. Watts (eds) *Globalising Food: Agrarian Questions and Global Restructuring*. London: Routledge, pp. 192–225.

Burch, D. and Goss, J. (1999) 'An end to fordist food? Economic crisis and the fast food sector in Southeast Asia'. In D. Burch, J. Goss and G. Lawrence (eds) *Restructuring*

Global and Regional Agricultures: Transformations in Australasian Agri-food Economies and Spaces. Aldershot: Ashgate, pp. 87–110.

Business a.m. (Edinburgh), 28 May 2002.

Business First of Louisville (Louisville, KY), 9 December 2002.

Business Journal of Jacksonville (Jacksonville, FL), 24 December 2002.

China Daily (Beijing), 14 March 2003.

Cox, A., Ireland, P., Lonsdale, C., Sanderson, J. and Watson, G. (2002) *Supply Chains, Markets and Power.* London: Routledge.

Department of Environment, Food and Rural Affairs (UK) (2004) 'Statistical data'. Online. http://statistics.defra.gov.uk/esg (accessed on 17 May 2004).

Dixon, J. (2002) *The Changing Chicken: Chooks, Cooks and Culinary Culture.* Sydney: University of New South Wales Press.

Dobson, P. (2003) 'Buyer power in food retailing: the European experience'. Unpublished paper presented to the *Conference on Changing Dimensions of the Food Economy: Exploring the Policy Issues,* 6–7 February, The Hague, Netherlands.

Farmers Weekly (Sutton), 6 February 2004.

Foreign Agricultural Service (FAS) (2002a) *India, Poultry and Poultry Products.* Washington (DC): United States Department of Agriculture.

Foreign Agricultural Service (FAS) (2002b) *Japan, Poultry and Poultry Products.* Washington (DC): United States Department of Agriculture.

Foreign Agricultural Service (FAS) (2002c) *Thailand, Poultry and Poultry Products.* Washington (DC): United States Department of Agriculture.

Foreign Agricultural Service (FAS) (2003a) *People's Republic of China, Poultry and Poultry Products.* Washington (DC): United States Department of Agriculture.

Foreign Agricultural Service (FAS) (2003b) *Hong Kong, Poultry and Poultry Products.* Washington (DC): United States Department of Agriculture.

Foreign Agricultural Service (FAS) (2004) 'Dairy, Livestock and Poultry Division, United States Department of Agriculture'. Online: http://www.fas.usda.gov/dlp/dlp.html (accessed on 18 May 2004).

Goss, J. (2002) *Fields of Inequality: The Waning of National Developmentalism and the Political Economy of Agribusiness: Case Studies if Development and Restructuring in Thailand's Agri-food Sector.* Unpublished PhD thesis. Brisbane: Griffith University.

Goss, J. and Burch, D. (2001) 'From agricultural modernisation to agri-food globalisation: the waning of national development in Thailand', *Third World Quarterly,* 22 (6): 969–87.

Goss, J., Burch, D. and Rickson, R.E. (2000) 'Agri-food restructuring and Third World transnationals: Thailand the CP Group and the Global Shrimp Industry', *World Development,* 28 (3): 513–31.

Grampian Country Food Group (GCFG) (2004) 'GCFG'. Online: www.gcfg.com (accessed on 24 January 2004).

The Guardian (London), 8 July 2002.

Handley, P. (2003) 'De-mythologizing Charoen Pokphand: an interpretive picture of the CP Group's growth and diversification'. In K.S. Jomo and B.C. Folk (eds) *Ethnic Business: Chinese Capitalism in Southeast Asia.* London, RoutledgeCurzon: 153–81.

Hazardous Waste Litigation Reporter, 21 November 2003.

The Herald (Glasgow), 3 April 2003.

Hughes, A. (1996) 'Forging new cultures of food-retailer relations?'. In N. Wrigley and M. Lowe (eds) *Retailing, Consumption and Capital: Towards the New Retail Geography.* London: Longman, pp. 90–115.

Johnson, D. and Turner, C. (2003) *International Business: Themes and Issues in the Modern Global Economy.* London: Routledge.

Lancashire Evening Post, 21 December 2001.

Memphis Business Journal (Memphis, TN), 6 December 2002.

Nation, (Bangkok), 28 October 1998; 11 May 2002.

Ngui, C.Y.K. (2001) 'The giant awakens', *Malaysian Business,* 1 December: 25.

Reuters News, 25 September 2002; 5 February 2004.

Stevens T., Hodges, A. and Mulkey, W. (2003) 'Economic impact of Tyson Food's plant closure in Northeast Florida, 2002'. Online: http://economicimpact.ifas.ufl.edu/com-pubs.htm (accessed 21 April 2004).

Sunday Herald (Glasgow), 21 March 2004.

Sunday Times (London), 4 April 2004.

Taiwan Economic News (Taipei), 21 May 2002.

Thai News Service (Bangkok), 6 November 1998; 12 February 2004.

Tyson Foods (2004) 'Tyson Foods, Inc., announces plan to consolidate Jackson and Carthage, Mississippi processing operations', press release, 27 February.

US Environmental Protection Agency (2003) 'Tyson Foods to Pay $7.5 million for Federal and State Clean Water Violation', press release, 3 July.

Vorley, B. (2001) *The Chains of Agriculture: Sustainability and the Restructuring of Agri-Food Markets.* Paper prepared for the World Summit on Sustainable Development. London: International Institute for Environment and Development.

Washington Business Journal (Washington, DC), 21 April 2003.

Wilkinson, J. (2002) 'Genetically modified organisms, organics and the contested construction of demand in the agro-food system', *International Journal of the Sociology of Agriculture and Food,* 10 (2): 3–11.

13 The difficulties of 'emerging markets'

Cross-continental investment in the South African dairy sector

Charles Mather and Bridget Kenny

Introduction

In the early 1990s transnational food corporations were identified as the key agents coordinating and driving cross-continental food systems (e.g. Bonanno *et al.* 1994; Heffernan and Constance 1994; Friedland 1994). Although large food companies had played a role in the global trade of food and fibre commodities for some time, usually associated with plantations of traditional crops such as coffee, tea and rubber (Dinham and Hines 1983), their role from the 1990s was seen as quantitatively and qualitatively different. These agents were arguably responsible for the coordination of global food chains in multiple production sites for rapid delivery to distanced consumption locales. While more recent research has tended to qualify the extent to which the food system is 'truly global' and comparable to globalized industrial production systems (Goodman and Watts 1994; Watts 1996), case studies of multinational food companies confirm that they continue to play an important role in the global agri-food system and in cross-continental food chains. Research on companies such as H.J. Heinz, the Charoen Pokphand Group, Nestlé, Cargill and ConAgra, among many others, has focused on the flexibility of sourcing practices, but also on the impact and response of producers in a wide range of local contexts (Heffernan and Constance 1994; Pritchard and Fagan 1999; Goss *et al.* 2000).

Much of the research on multinational corporations in the global food system focuses on how these organizations source products, often but not exclusively in developing countries, and then supply this food in a fresh or processed state to wealthy consumers in developed countries. Since the late 1980s, however, there has been considerable direct investment by multinational companies to service *domestic markets*. This investment has, for obvious reasons, concentrated on developed market economies, but it has also occurred in 'emerging market' countries where rising incomes have seen rapid changes in consumption patterns and new demands for non-staple products, mainly dairy, meat and fresh fruits and vegetables. The sale or processing of food commodities by multinationals to supply domestic markets seems counterintuitive given that trade barriers, especially in emerging market economies, are dramatically lower as a result of market liberalization. Yet for highly perishable commodities such as dairy, supplying domestic markets demands investment in local processing facilities, which has occurred chiefly through mergers or acquisitions by multinational dairy companies. There is now a growing body of work that has explored the acquisition of processing and retailing capacity by multinational firms, and the impact this has had on competition, primary production and consumption (Driven 2001; Faigenbaum 2002; Gutman 2002).

This chapter examines these issues via a case study of recent investments in South Africa by Parmalat and Danone, two dairy multinationals of Italian and French parentage respectively. For the wider debate on cross-continental food chains, this case study brings into focus three important points. First, it emphasizes the role of investment, as opposed to trade, in the international integration of food systems. Second, it highlights the highly competitive local economic landscapes that can confront multinational companies in emerging markets. And third, it contributes to recent scholarship that documents shifts in market power along the food chain, from processing companies to supermarket retailers. Material herein is derived from industry published and unpublished sources, and interviews with dairy farmers, processors and retailers.[1]

With regard to the local impacts of these investments, reference needs to be made to the geographically variable and culturally specific characteristics of dairy consumption (Pritchard 2002). South African dairy consumption patterns do not mirror those of other emerging markets where urbanization, higher incomes and the spread of formal retailers have led to rapid increases in milk consumption (Faiguenbaum *et al.* 2002; Gutman 2002; Reardon and Berdegue 2002). In South Africa, dairy consumption continues to be shaped by the legacy of apartheid. Per capita consumption of dairy products has declined from 65 litres in 1989 to only 43 litres in 2000. Only 12 per cent of the population drinks milk on its own; 2 per cent use butter, and between 6 per cent and 8 per cent eat cheese. These conditions reflect South Africa's economic crisis (associated with massive job losses and sharp declines in disposable income), weak efforts to encourage dairy consumption among African households (it has been estimated that 'non-black' households, mostly middle-class and upper-class whites, account for up to 70 per cent of national dairy consumption: Yankelevich 1999) and competition from non-dairy substitutes (creamers and margarine), aided and abetted by weak legislation on labelling.[2] There is, however, evidence that the consumption of some processed dairy products, including yoghurt, drinking yoghurt and ultra-high temperature (UHT) long-life milk, is increasing, primarily as a result of new product innovation and promotions.

Transforming the competitive space of processing

In the late 1990s Parmalat, the Italian multinational, and Danone, the French dairy giant well known for yoghurt products, made significant investments in the South African dairy industry. While Danone established a partnership with the country's oldest cooperative (National Cooperative Dairies, or 'Clover'), Parmalat was decidedly more aggressive and purchased a large privatized dairy cooperative called Bonnita and a small dairy cooperative called Towerkop. These investments came at a time of considerable crisis in the processing sector, largely as a consequence of the challenges posed by liberalization. Both of Parmalat's acquisitions were in considerable financial difficulty when the offers to purchase were made. Similarly, Danone's partnership with Clover has played an important role in ensuring the financial viability of the country's largest and oldest dairy cooperative.

As in other parts of the world South Africa's dairy industry was strongly regulated. The Dairy Board set producer and retail prices for milk, butter and other dairy products; there were controls on the registration of processors and their supply regions; and levies were imposed on farmers and milk processors to fund what was called the 'surplus removal scheme'. That scheme had two main functions: it prevented 'unnecessary' competition between processors and it protected processors from cyclical changes in the supply of milk by storing or exporting surplus product, usually at a loss. From the mid-1980s the dairy sector was liberalized: price controls were relaxed and later removed, and after 1988, restrictions on the establishment of processing facilities were lifted. The lifting of the regulations governing the number of processors had an immediate impact on the structure of the processing sector. In 1987 there were around 40 processors; by 1994 the number had increased to over 500. On the one hand, this explosion in the number of processors did not dilute concentration significantly, with the five largest processors continuing to control about 70 per cent of fresh milk production. Yet it nevertheless transformed the competitive landscape of the sector. Medium-sized new entrants were not burdened by the large overhead costs associated with their larger competitors supplying the national market, enabling them to supply fresh milk at prices 10–30 per cent below the large processors. Smaller processor-distributors – who usually sell through small 'milk shops' in large cities – could sell milk at an even lower prices, although there have been reports of quality problems from some of these operators (More O'Ferrall-Berndt 2003).

The new competitive structure of dairy processing had an unintended consequence of further liberalizing the processing sector. The existence of many new processors compromised the ability of the Dairy Board to manage the surplus removal scheme, which effectively protected processors from cyclical changes in the supply of milk. By the early 1990s the Dairy Board's debt was such that it was forced to abandon the surplus removal scheme. From this period on, processors were no longer protected from seasonal and cyclical changes in milk production.

From the early 1990s large and medium-sized dairy processors found themselves having to deal with the twin problems of intense competition from low-cost processor-distributors and cyclical changes in the supply of fresh milk. Bonnita and Towerkop responded to increased competition by developing new long-life product lines, including yoghurt, cheese, ice cream and UHT milk. This diversification reduced the two processors' exposure to the competitive pressures of fresh milk, while at the same time provided partial protection from the vagaries of milk supply.

Towerkop's forays into ice cream and yoghurt did not enable the cooperative to escape financial danger, whereas Bonnita fared far better. An amendment to the Cooperatives Act in 1993 allowed Bonnita to shed its cooperative status and become a private company with shares held by farmers. Shortly afterwards, Bonnita secured a R110 million (US$17 million) investment from Premier Foods, a large and highly diversified food company, and listed on the Johannesburg Stock Exchange. The company grew rapidly after its listing. Following its 1995 results Bonnita was praised for its ability to weather milk 'production overruns', a 'chronic problem' facing the dairy sector (*Finance Week* 1995). The financial press was also impressed by its aggressive regional strategy through exports and its purchase of the Zambian Dairy Cooperative. And despite huge investments it remained relatively debt free. As a result the company's share price almost doubled in two years. By 1997, however, reports on Bonnita were much less favourable. Although the company had weathered a cyclical oversupply of milk in the mid-1990s, it was much less successful in dealing with severe shortages in the 1997 season. Higher producer prices and competition from smaller processors were having a serious impact on earnings. Rumours soon spread that its holding company, Premier, planned to unburden itself of a company that had successfully transformed itself from a cooperative to a private company but that nonetheless remained vulnerable to the vagaries of milk supply and intense competition in the processing sector.

Parmalat purchased Bonnita from Premier Foods in 1998, and a year later the Italian company also acquired Towerkop. Parmalat's decision to purchase these two processors was almost certainly an attempt to control the supply and market for milk in the Western and Eastern Cape regions of the country. Although the processing sector had been deregulated for several years, a legacy of the regulated sector was a North–South divide for both sourcing and supplying milk products. By purchasing Towerkop and Bonnita, Parmalat secured the second largest market for dairy products in South Africa (Western Cape) and a virtual – but short lived – monopsony over the supply of fresh milk in the Western and Eastern Cape.

South Africa's largest dairy processor, Clover, also struggled in the face of competition from other medium-sized and small dairy companies. Competition in the area around Johannesburg and Pretoria, the country's largest market for dairy products, has been intense in the period since the liberalization of the dairy sector: many of the small processor-distributors exist close to these major urban centres. Clover's strategy since the early 1990s has been to upgrade its production and distribution facilities and to buy up smaller dairy cooperatives. However, these two strategies have been expensive, and by the mid-1990s Clover's debt was estimated at well over R400 million (US $61.5 million). At the same time, the cooperative was losing market share and facing lower margins on fresh milk in the new competitive environment. Despite a massive turnover of almost R2 billion (US $335 million) in 1995, its profit was less than R31 million (US $4.8 million), mainly because of debt servicing. Thus in 1996, when it was announced that Danone, the French multinational dairy company, planned to invest R400 million (US $61.5 million) in Clover, it seemed that the domestic processor's key challenges would be solved. The funds would allow the company to pay off its crippling debt, and Danone's expertise in yoghurt would be used to develop new fresh and processed dairy products. However, agreement on the terms of the investment dragged on for two years due to 'tricky management and culture issues' (Reid 2000: 20) and were only concluded in late 1998. The investment resulted in the establishment of a company called Danone Clover, focused exclusively on fermented dairy products such as yoghurt and drinking yoghurt.

Notwithstanding these investments by the two multinationals, the South African dairy sector has proved to be a stubborn terrain from which to make profits. It has remained vulnerable to competition from low-cost processor-distributors and to cyclical changes in milk supply. The companies purchased by Parmalat – Bonnita and Towerkop – had operated almost exclusively in the Western and Eastern Cape areas of the country, where dairy production has increased rapidly in the last decade (see Figure 13.1). Parmalat's acquisition of the two dairy processors coincided with a cyclical upturn in milk production and the Italian processor soon found itself with too much milk. An announcement in 1999 that the company could not take up to 10 per cent of the milk produced by its farmer suppliers resulted in a mass defection of dairy producers to a medium-sized competitor. In response, Parmalat was forced to offer large dairy producers a price increase of up to 5 per cent per litre of milk, based on a three-year supply contract.

Clover-Danone has faced equally difficult challenges. Despite the multinational's huge cash infusion, the company faced bankruptcy in 2002. Unwilling to see its initial investment collapse, Danone has made a further large, but undisclosed, investment. This new investment is closely aligned with a significant restructuring of Clover, which in turn reflects some of the key challenges facing dairy processors. Danone's relationship with Clover will now be restricted to parts of the company that produce yoghurt, drinking yoghurt, chocolate and other flavoured-milk products. These products are regarded as less vulnerable to problems in the supply of fresh milk and can also be strongly branded and promoted through advertising campaigns.

The profit pressure on dairy processors has led to dramatic changes in the geography and structure of dairy farming in South Africa. Over the last decade there has been a significant shift of production towards coastal areas where it is possible to farm using cheaper pasture-based systems (Figure 13.1). Inland farmers tend to rely on maize-based feed systems, the costs of which have risen sharply following the weakening of the South African currency. The shift has been most dramatic in the Eastern Cape: in the late 1990s this region produced around 13 per cent of total production; the most recent figures show that the region now produces over 20 per cent of the country's total milk supply. Further, the number of dairy farmers has declined sharply, especially in recent years. In December 1997 there were almost 8,000 dairy farmers, but by 2002 there was estimated to be just over 4,000. Moreover, declines in the number of dairy farmers have also occurred in areas where milk production has been rising. There were over 700 dairy producers in the Eastern Cape in 1997, but only 480 in 2002. At the same time, not surprisingly, the average size of dairy farms has increased, and the larger farms now have a much higher percentage of total production (Table 13.1). Farmers with a daily production of more than 6,000 litres now contribute 24 per cent of total production, up from 10 per cent in 1995.

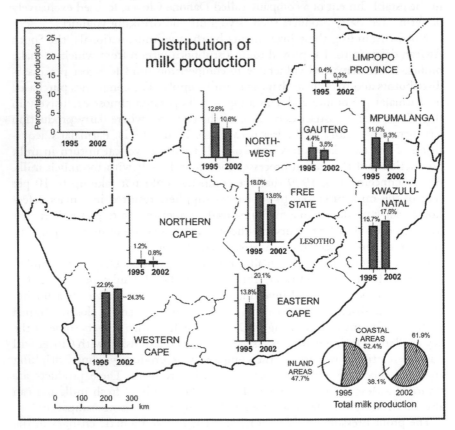

Figure 13.1 The changing geography of dairy production in South Africa
Source: Coetzee 2003.

Table 13.1 Structure of dairy farming in South Africa

Daily production (litres per day)	Percentage of farm enterprises		Percentage of production	
	1995	2001	1995	2001
0–500	58	45	19	9
501–1,000	21	17	20	9
1,001–2,000	13	17	24	19
2,001–4,000	6	11	22	24
4,001–6,000	2	5	5	15
6,000+	0	5	10	24

Source: Coetzee (2003).

Large and medium-sized processors have played a key role in driving these changes, as they have done in other parts of the world (Banks and Marsden 1997; Breathnach 2000). Previously, industry cost efficiencies were impaired by the existence of many small, and usually part-time, dairy farmers. Collecting milk from these farmers constituted an enormous financial burden for processors. They have, as a result, provided incentives for dairy farmers to produce more milk. All South African based processors now pay a premium on higher volumes of milk, usually 1 cent per litre for every 1,000 litres. Processors have also rationalized collection routes and smaller farmers 'off the beaten track' are now finding it very difficult to secure a buyer for their milk. New quality requirements, which usually require significant capital investments, have placed further pressure on smaller dairy producers. Finally, it seems very likely that processors have kept producer prices low enough to force smaller dairy farmers to sell their cows to larger farmers. The impact of 80 farmers protesting low producer prices in August 2002 was largely ignored by processors because their total production of milk was only 80,000 litres per month. Processors recommend that dairy farmers increase their herd size to more than 200 or 'discontinue milk production altogether' (Yankelevich 1999).

Retailing dairy

Retailers have taken advantage of the competitive pressures in the processing sector by squeezing prices, especially in fresh milk where the competition is most intense. South African food retailing is a highly concentrated and saturated market. As in other parts of the world (e.g. Wrigley and Lowe 1996, 2002), corporate retail chains have increased their market share and have consolidated power over processors in the supply chain. South African food retailing is characterized by a handful of large national chains with great buying and bargaining power. Although smaller shops and informal traders are estimated to account for approximately 30 per cent of total retail spending (Thomas 2003), this sector tends to be geographically differentiated from the larger urban-based markets. In 2002, South Africans were estimated to have spent US$6 billion on food through formal commercial outlets, and an estimated 55–60 per cent of this was made through the major supermarkets and hypermarkets (Weatherspoon and Reardon 2003). This figure is comparable to the share of supermarkets in Argentina, Chile, the Philippines and Mexico, but behind that of the United States (at 70 per cent) (Weatherspoon and Reardon 2003). This concentration in ownership gives powers to corporate food retailers to define the character of domestic food markets.

Four companies control 95 per cent of the supermarket sector: Shoprite/Checkers, Pick 'n Pay, Spar and Woolworths. This represents a high level of concentration when compared with other countries (for example, in the UK five companies account for just over 50 per cent of the market: Hughes 1996). The four major South African supermarket chains each specialize, so Shoprite vies for the lower-income market with increasing competition from Pick 'n Pay; Spar, Pick 'n Pay and Shoprite's Checkers brand compete for the middle- and upper-income markets, and Woolworths serves the upper-income market.

Power relations between retailers and dairy processors are decidedly in favour of the former. Retailers exercise control over their suppliers through 'preferred supplier' agreements (cf. Wrigley and Lowe 2002). Preferred suppliers are 'listed' with corporate retailers, a process that requires meeting the retailers' specific criteria (Weatherspoon and Reardon 2003). Although the listing process requires an upfront payment, more significant is the supplier's ability to meet the product and delivery capabilities, and price. Clover-Danone and Parmalat are listed at all the major retailers. Medium-sized processors may be listed with one or two retailers, but seldom at more. Processor-distributors tend not to be listed by supermarket retailers, as generally they cannot meet volume and quality requirements.

On the other hand, the smaller franchise outlets of the major retailers are usually more flexible in their choice of dairy supplier. Because they compete with corner stores that are supplied with cut-price milk by processor-distributors, they are often more amenable to being supplied by cheaper, smaller processors. Therefore, the fragmented structure of processing has played into the hands of retailers. For supermarkets, medium-sized dairies with lower overhead costs are used as a competitive lever against larger processors. In smaller franchise stores, processor-distributors are used to place competitive pressure on both medium- and larger-scale processors.

Retailers further squeeze suppliers through regularly negotiated discounts and rebates, as well as charging suppliers extra costs for promotions, new product listings and returns for milk that has passed its 'sell-by' date. The absence of contracts between retailers and processors allows the former to negotiate on price and discounts. Retail buyers normally negotiate a price per litre for milk and then demand a 'rebate' based on the volume of supply; as a general rule rebates are higher for larger volumes of milk. Besides a standard discount, retailers also demand from processors periodic price cuts that are blamed on the competitive pressures in the retail sector or on seasonal oversupplies of milk. During negotiations retailer milk buyers may provide evidence, through advertisements, on the prices of their competitors and they use this to demand further price cuts from processors. As one processor rhetorically asked the current researchers: 'Have you seen a milk buyer's office? They have all their competitor's milk prices stuck on their pin-up boards.' Interviews with processors confirmed that most felt that they had little choice but to submit to retailer demands. In one case, a processor was de-listed for refusing to accept a rebate of R1 (US 60 cents), equivalent to about 25 per cent of the price of milk. Demands for discounts are not restricted to smaller processors; even the largest processors such as Clover-Danone and Parmalat are not immune from the pressures of retailer power.

Additionally, listed processors are paid for dairy products delivered on a credit cycle of between 30 and 90 days. According to one report, delays in payment form a significant 'profit source' for retailers:

> They [retailers] get money from interest. We're lucky if we get our money 45 to 60 days after delivery. But their customers pay cash. They can also delay paying VAT [value-added tax] by another 60 days. Supermarkets aren't concerned about price. They want turnover so they can put the money through the bank.
>
> (*Cape Times* 1997)

According to some processors interviewed for this research, retailers may also delay payments for minor errors on invoices and other small administrative problems. As one processor suggested: 'Pick 'n Pay's payment terms are 60 days. But if there is a 1[cent] mistake on your R20 million (US $3.1 million) invoice, you will have to wait another month for your money.'

Returning dairy products that have passed their 'sell-by' date represents another cost for processors. Retailers return unused milk that has passed its 'sell-by' date to suppliers at full reimbursement (they also do this for spoiled product). According to one processor, the average return is between 5 and 6 per cent by volume and these costs are incurred for all processors regardless of size. Processors we interviewed complained that retailers were not willing to consider discounting dairy products that were close to reaching the 'sell-by' date.

The pressures placed on dairy processors through the South African retail environment have provided a further justification for diversification. In recent years, Parmalat and Clover-Danone have moved into value-added products, which are heavily promoted and strongly branded. These efforts have resulted in dynamic changes in the drinking yoghurt, yoghurt, chocolate milk and dairy-juice products market. Indeed, processor efforts appear to have increased consumption of these products in the context of dramatic declines in overall dairy consumption. These activities must be understood primarily as an attempt to leverage competitive space with retailers. However, the success of these endeavours is often compromised by the fact that retailers respond quickly to innovations by developing similar own-label products. In similar fashion to processes observed internationally (Doel 1996), South African retailers appear to use own-labels to lower processors' prices and to increase their bargaining power. Retailers also use the awarding of contracts to supply own-label brands to enhance competition among processors. One retailer interviewed for the current research noted: 'We didn't want to give the contract [for the own-label product] to Clover because they are so big. They were bitter.'

Conclusion: South Africa's changing agri-food landscape

The difficulties faced by Parmalat and Danone highlight the risks for multi-national corporations that invest in emerging markets to service domestic markets. While the 'model' of emerging market consumption patterns suggests that political reform and economic growth usually result in the rapid increases in the consumption of non-staple commodities (cf. Farina and dos Santos Viegas 2003), the South African case suggests this may not always be the case. Dairy consumption has declined rapidly through the 1990s and shows little sign of recovery. Although there is some evidence that the two organizations have increased the consumption of new dairy products – especially yoghurt and drinking yoghurt – this has occurred within the narrow confines of a small middle-class white market. The inability of the two multinational processors to transform South Africa's dairy market reflects the extent to which markets are not 'universally dominant' or 'inevitable' and are 'shaped by multiple histories, geographies and culture' (Greenaway *et al.* 2002: 720). In South Africa's case, the two corporations have been unable to overturn the situation whereby the small white minority consume almost 70 per cent of the country's dairy products sold in formal retail outlets.

The problems Parmalat and Danone faced in South Africa's dairy market were compounded by changes in the country's agri-food landscape. In the last decade, food retailing has become highly concentrated with four companies controlling most food sales. Not surprisingly, they wield consid-erable power over their suppliers, who are now subject to a range of buying practices that place pressure on their margins. At the same time, the liberal-ization of agricultural markets in South Africa, which made the investments by Parmalat and Danone possible, has resulted in intense competition in the processing sector. In order to secure shelf-space in this highly competitive environment, processors have had to continually present new products that are heavily promoted in the print and electronic media. This situation is not unique to South Africa: commenting on food multinationals in the Brazilian economy, Farina and dos Santos Viegas (2003: 14) note that 'enter-ing alone does not guarantee multinationals a comfortable situation'. The uncomfortable situation facing food multinationals in Brazil, as in the case of South Africa, is associated with a liberalized market environment, intense competition in the processing sector, and a large and powerful retail sector.

The difficulties associated with emerging markets may be part of the explanation for the financial scandal and bankruptcy of Parmalat revealed in late 2003. Although the cause of the organization's financial collapse is multiple and complex, financial analysts now suggest that its strategies in emerging markets were 'reckless' and lacked a proper 'marketing strategy' (Colitt 2004). As a result, many of the multinational's foreign affiliates experienced heavy financial losses, including Brazil where the company had not generated a profit since 1997. Not surprisingly, one of the restructuring options for Parmalat involves selling all of its foreign affiliates and refocusing its efforts in Italy and then perhaps Europe. If the Parmalat case reflects a much broader process of 'misreading' emerging markets – and not simply a case of financial fraud – its collapse may be far more significant to the long-term process of cross-continental investment in the agri-food sector.

Notes

1   This research is based on interviews conducted during June and July 2003 with six large and medium-sized dairy processors based around Johannesburg and Pretoria. Our material on retailers is based on interviews with dairy buyers from large supermarkets. We also interviewed two dairy processors and fifteen dairy farmers in the Eastern Cape during August 2003.
2   For instance, companies are permitted to produce and brand 'ice-cream' using various non-dairy products including vegetable fats.

References

Banks, J. and Marsden, T. (1997) 'Reregulating the UK dairy industry: the changing nature of competitive space', *Sociologica Ruralis*, 37 (3): 382–404.
Bonnano, A., Busch, L., Friedland, W., Gouveia, L. and Mingione, E. (eds) (1994) *From Columbus to ConAgra: The Globalization of Agriculture and Food*. Lawrence (KS): University of Kansas Press.
Breathnach, P. (2000) 'The evolution of the spatial structure of the Irish dairy processing industry', *Irish Geography*, 33 (2): 166–84.
Cape Times (1997) 'Dairy demands huge increases', 17 June, p. 6.
Coetzee, K. (2003) *Lacto Data*, Milk Producers' Organisation, Pretoria, South Africa.
Colitt, R. (2004) 'Parmalat pays high price to conquer Brazil', *Financial Times*, 6 January 2004.
Dinham, B. and Hines, C. (1983) *Agribusiness in Africa*. London: Earth Resources Research.
Doel, C. (1996) 'Market development and organizational change: the case of the food industry'. In N. Wrigley and M. Lowe (eds) *Retailing, Consumption and Capital: Towards the New Retail Geography*. Harlow, Essex: Longman, pp. 48–67.
Driven, M. (2001) 'Dairy clusters in Latin America in the context of globalization', *International Food and Agribusiness Management Review*, 2 (3/4), 301–13.
Faigenbaum, S., Berdegue, J.A. and Reardon, T. (2002) 'The rapid rise of supermarkets in Chile: effects on dairy, vegetable and beef chains', *Development Policy Review*, 20 (4): 459–71.
Farina, E.M.M.Q. and dos Santos Viegas, C.A. (2003) 'Multinational firms in the Brazilian food industry'. Unpublished paper presented at the International Food and Agribusiness Management Association Conference, Cancun, Mexico.

Finance Week (1995) 'Bonnita: Go for long shelf life', 17, 8–14 June.

Friedland, W.H. (1994) 'The new globalization: the case of fresh produce'. In A. Bonnano, L. Busch, W. Friedland, L. Gouveia and E. Mingione (eds) *From Columbus to ConAgra: The Globalization of Agriculture and Food*. Lawrence (KS): University of Kansas Press, pp. 210–30.

Goodman, D. and Watts, M. (1994) 'Reconfiguring the rural or fording the divide? Capitalist restructuring and the global agro-food system', *Journal of Peasant Studies*, 22 (1), 1–49.

Goss, J., Burch, D. and Rickson, R.E. (2000) 'Agri-food restructuring and third world transnationals: Thailand, the CP group and the global shrimp industry', *World Development*, 28 (3), 513–30.

Greenaway, A., Larner, W. and Le Heron, R (2002) 'Reconstituting motherhood: milk powder marketing in Sri Lanka', *Environment and Planning D: Society and Space*, 20 (6) 719–36.

Gutman, G.E. (2002) 'Impact of the rapid rise of supermarkets on dairy product systems in Argentina', *Development Policy Review*, 20 (4): 409–27.

Heffernan, W.D. and Constance, D.H. (1994) 'Transnational corporations and the globalization of the food system'. In A. Bonnano, L. Busch, W. Friedland, L. Gouveia and E. Mingione (eds) *From Columbus to ConAgra: the Globalization of Agriculture and Food*. Lawrence (KS): University of Kansas Press, pp. 29–49.

Hughes, A. (1996) 'Forging new cultures of food retailer–manufacturer relations?'. In N. Wrigley and M. Lowe (eds) *Retailing, Consumption and Capital: Towards the New Retail Geography*. Harlow, Essex: Longman, pp. 90–115.

More O'Ferrall-Berndt, M. (2003) 'A comparison of selected public health criteria in milk from milk–shops and from a national distributor', *Journal of the South African Veterinary Association*, 74 (2): 1–12.

Pritchard, B. (2002) 'Current global trends in the dairy industry'. Unpublished presentation to the International Union of Foodworkers' Global Dairy Conference, Auckland (New Zealand). Available from the author: Division of Geography, University of Sydney 2006 Australia.

Pritchard, B. and Fagan, R.H. (1999) 'Circuits of capital and transnational corporate spatial behaviour: Nestlé in Southeast Asia', *International Journal of Sociology of Food and Agriculture*, 8: 3–20.

Reardon, T. and Berdegue, J.A. (2002) 'The rapid rise of supermarkets in Latin America: challenges and opportunities for development', *Development Policy Review*, 20 (4): 371–88.

Reid, R. (2000) 'Milk – but little honey (South African dairy industry)', *Dairy Industries International*, January, 65 (1), 20.

Thomas, S. (2003) 'Food retailers: no longer such a safe bet', *Financial Mail* (South Africa), 170, 44.

Watts, M. (1996) 'Development III: the global agrofood system and late twentieth-century development (or Kautsky *redux*)', *Progress in Human Geography* 20, 230–45.

Weatherspoon, D. and Reardon, T. (2003) 'The rise of supermarkets in Africa: implications for agrifood systems and the rural poor', *Development Policy Review*, 21 (3), 333–55.

Wrigley, N. and Lowe, M. (eds) (1996) *Retailing, Consumption and Capital: Towards the New Retail Geography*. Harlow, Essex: Longman.

Wrigley, N. and Lowe, M. (2002) *Reading Retail: A Geographical Perspective on Retailing and Consumption Spaces*. London: Arnold Press.

Yankelevich, I. (1999) *Industry Overview: The Dairy Industry*. Johannesburg: Standard Bank of South Africa Economic Research Sectoral Research Unit.

Part IV

Multi-scalar politics and the restructuring of cross-continental food chains

14 The politics of place

Geographical identities along the coffee supply chain from Toraja to Tokyo

Jeffrey Neilson

Introduction

There exists a widespread industry belief that coffee grown in a specific geographic region will retain certain quality attributes reflecting the physical and human characteristics of that growing environment. As with other selected food and beverage products, such as wines, cheeses and meats, the geographic place name of the producing region, or nearby trading centre, is subsequently applied as a marketing or trade identity by various actors along the commodity chain to indicate the presence of these quality attributes. This chapter explores the use of geographical identities along a coffee supply chain extending from rural origins in the Toraja region of Sulawesi in eastern Indonesia through to consumption in cosmopolitan Japan.[1] The economic importance, appropriation and legal protection of the 'Toraja' identity in the supply chain make this a particularly informative case study. Sulawesi coffee is consumed predominantly as a single-origin offering in Japan, where roaster-retailers emphasize regionality to obtain the substantial price premium afforded to this place-informed gourmet product. Geographical specificity and control over the use of the 'Toraja' identity have become critical determinants of economic relationships between supply chain actors.

Supply chain regulation through the consumption of place

In the early colonial development of the global coffee industry, geographical identities were an integral element of the descriptive language employed by traders and roasters, who promoted romantic images of the exotic locations where coffee was grown. Decolonization, together with advances in roasting and brewing technology, gradually eroded the importance of regional agricultural associations, as control of product identity shifted to the roasting sector, which was invariably located at the site of consumption. Pendergrast (2001) describes how large-scale roasters completed the transformation of coffee away from a place-differentiated product into one where green beans were used largely as an undifferentiated, commoditized input for industrial processing. The declining use of regional coffee identities coincided with a period of tightly regulated international trade implemented periodically by the International Coffee Organization (ICO) between 1962 and 1989. Rigid control of national outputs through an international quota system during this period seemed to contribute to the standardization of coffee quality associations. More recently, however, a renewed interest in the regional origins of coffee products has emerged in major consuming countries associated with strong growth in the specialty coffee sector.

The specialty coffee industry is now the most vibrant and fastest-growing segment of the global coffee market, although the actual definition of specialty coffee continues to be debated (Specialty Coffee Association of America (SCAA) 1999; ICO *et al.* 2000; International Trade Centre (ITC) 2002; Ponte 2002a; Lewin *et al.* 2004; SCAA 2004). The role of geography in influencing taste profiles and the preservation of geographical identities in the coffee trade were central concerns for Norwegian coffee connoisseur Erna Knutsen, who is widely accredited with coining the term 'specialty coffee' in the 1970s (Holly 2003). The founding members of the Specialty Coffee Association of America (SCAA) similarly agreed to define specialty coffee in 1982 as 'good preparation from unique origin and distinctive taste' (cited in Ponte 2002b: 11). The proliferation and popularity of roaster-retailer chains, offering coffee beans from various regional origins, provide further support for the increased prevalence of geographical specificity within the specialty sector. The explicit importance of geographical influences in the construction of specialty coffee however, remains polemical. Specialty roasting companies rely heavily on their own marketing image and product branding. The widespread use of milk, sugar and other flavourings in espresso bar culture is also often at the expense of an emphasis on agricultural origins. Despite such ambiguities, the use of geographical identities appears to have increased within roasted coffee marketing over the last 15 years.

The quality associations of particular place names are commonly presented as a function of how production is embedded within geographic space. Granovetter's (1985: 482) theory of embeddedness argues that economic behaviour is 'constrained by ongoing social relations'. In this chapter, embeddedness is considered in its geographical context, where the totality of place-specific socio-cultural, economic and environmental influences interacts with coffee production in a way which significantly affects the dynamics of quality construction. The notion of geographical embeddedness is thus used to describe the entanglement of place and quality construction at the site of agricultural production. Importantly, however, the quality associations of particular forms of geographical embeddedness are mediated by, and translated to consumers through, vertically oriented supply chain structures. Through these structures, any inherent value associated with the nature of geographical embeddedness can be manipulated, controlled and reconstructed by non-local, and therefore geographically 'disembedded', actors.

The potential of specialty food products utilizing regional identities to provide an alternative development approach for lagging rural regions has been discussed by Ilbery and Kneafsey (1999) and Ray (2001). Implicit to these analyses is the ability of producers to capture the economic value of quality linked to the geographical embeddedness of production. Indeed, consumer preferences for geographically specific and regional food products are frequently set within the context of emerging regulatory structures associated with the protection of collective intellectual property through appellations systems and Protected Geographical Indications (Moran 1993; Parrot et al. 2002; Barham 2003). The implementation of such protection, however, requires substantial financial support from public institutions and highly specific legislative arrangements. In the context of global supply chains, this local support must be further sustained by recognition and a shared understanding of international laws and trade agreements. For the coffee supply chain extending from Sulawesi to Japan, corporate regulation of the use of a regional identity has emerged in the absence of such producer-driven collective protection. This outcome has important implications for market access and the distribution of economic benefits associated with the qualities of geographical embeddedness.

Sulawesi coffee production and export identities

During the period 1991 to 2003, an average 2,500 tonnes of *arabica* coffee were exported annually from the port of Makassar in South Sulawesi, accounting for less than 10 per cent of Indonesia's total *arabica* exports.[2] However, for the highland communities where coffee is grown, its production is the principal source of cash income. In Tana Toraja district for example (Figure 14.1) coffee contributes an estimated 20 per cent to the gross domestic regional product in an economy otherwise dominated by subsistence rice production (Badan Pusak Statistik (BPS) 2002). Nearly all Sulawesi coffee is sold by exporters to specialized green bean traders in the US, Japan and Europe, who then supply the growing specialty coffee sector in those consuming regions. All exports of *arabica* coffee from Sulawesi in 2002 and 2003 were marked, and subsequently traded internationally, using a geographical place name such as 'Toraja', 'Kalosi', 'Rantepao', 'Sulawesi' or 'Celebes', and all were designated Grade One export quality. These identities feature on the ICO Certificate of Origin, related export documentation, and are commonly printed on individual 60-kilogram bags of coffee. The identity therefore remains the principal means of product differentiation up to the point of roasting. At this point, roasted single-origin coffee is commonly sold under this same geographic identity, or alternatively may assume another identity depending on marketing priorities.

Prior to Dutch colonization of the Toraja highlands in 1905, coffee was traded through indigenous networks west to the port town of Boengie (Figure 14.1), from where it was shipped to the main export hub of Makassar

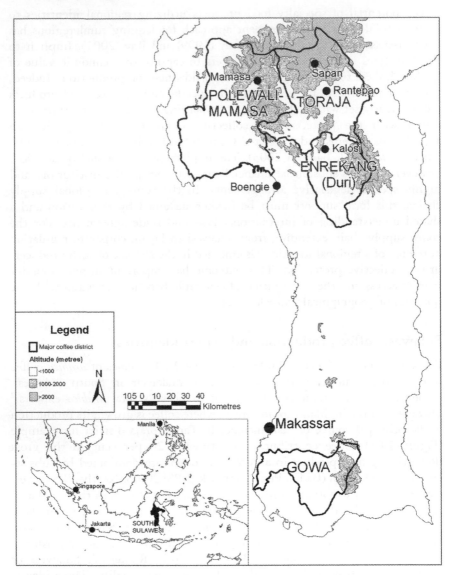

Figure 14.1 South Sulawesi coffee-growing areas

Source: Author.

(Bigalke 1981). This coffee was then known as Boengie coffee, and had a particularly good reputation in the European market (Ukers 1935). Smallholder growers belonging to the Duri and Toraja ethnic groups cultivated this coffee, with Duri traders from the town of Kalosi assuming an important role in local trade networks at this time. Even after trade networks were no longer conducted through Boengie port, the Dutch administration recognized the value contained within a place name, and actively attempted to 'protect the good name of Boengie coffee' (Paerels 1949: 106) throughout the colonial period.

Following the Indonesian declaration of independence in 1945, regional instability across South Sulawesi culminated in violent ethno-religious tensions between the Duri and Toraja ethnic groups, disrupting inter-regional coffee trade networks. Very little Toraja-grown coffee reached the export market during this period because of the risks associated with trans-port along the volatile trade routes (Bigalke 1981). Kalosi traders, collecting coffee grown predominantly in the Duri lands of Enrekang, were apparently responsible for the widespread use of Kalosi as a market identity at this time. Interestingly, both the Boengie and Kalosi identities were borrowed from trading hubs and did not refer to a specific growing region.

While Kalosi continues to be a popular trading name today, Table 14.1 indicates that during 2002, Toraja had emerged as the most popular geo-graphical expression used to identify exports of *arabica* coffee from Makassar. The contemporary use of geographical identities, however, demonstrates significant variation dependent on export destination. 'Kalosi' is primarily used for European buyers, while each of the 'Toraja' and 'Kalosi' identifiers are popular for the US market, and the Japanese indicate a clear preference for 'Toraja'. In the Sulawesi coffee supply chains, it is common practice for exporters to apply whichever identity is requested by international buyers. These export identities therefore currently reflect consumer market recogni-tion, rather than necessarily corresponding to actual coffee origin. Moreover, importers have been known to deliberately mislabel Sulawesi coffee as Mandheling, the well-known coffee-growing region in North Sumatra.[3] Mandheling was the third most popular geographical expression used for Makassar exports in 2002 (Table 14.1). While many industry actors recognize and promote the quality associations of particular growing

Table 14.1 Geographical identities for *arabica* coffee exports from Makassar Port, 2002 (%)

Geographical identity	Export destination				
	Europe	US	Japan	Other Asia	Total exports
Toraja only	0	23	90	20	41
Kalosi only	64	23	3	11	22
Both Toraja and Kalosi	21	25	6	47	20
Mandheling	9	28	0	21	15
Other	6	1	0	1	2
Total exports (tonnes)	574	1,613	1,194	263	3,644

Source: ICO Certificates of Origin, sighted at Department of Industry and Trade, South Sulawesi.

environments, the authenticity of the use of geographic identities is, for the most part, poorly regulated.

Toraja is a growing region and not a trading centre. The increased popularity of Toraja as an export identity therefore appears to represent a heightened international appreciation of the embedded geographies of coffee production in Sulawesi. Toraja is one of the four principal coffee-growing districts in South Sulawesi (Figure 14.1). These four districts are inhabited by distinct ethnic groups, each with a unique language, cultural practices and agro-ecological systems into which coffee production has been inserted. The physical characteristics of each growing district differ significantly in terms of topography and altitude, weather patterns and soil types. On-farm processing of the coffee bean also reflects these divergent geographies, with harvesting, pulping, fermentation and drying methods ranging widely. Of the Sulawesi growing districts, the highest-quality coffee is widely considered by local traders, exporters and international buyers to be grown in the Toraja district. The quality associations of Toraja coffee are primarily attributed to higher altitude production, along with prevailing soils, cultural characteristics[4] and to local processing methods. However, the ability of most international buyers to trace accurately the local origins of Sulawesi coffee is severely limited by the complex nature of existing pre-export trading systems.

The politics of place

Japan's second largest coffee company, Kimura Coffee (later to become Key Coffee), established a coffee plantation and processing mill in the Toraja region during 1977. According to company pioneers in Toraja, the industry was then in a state of disarray, with little coffee reaching the export market and plantations in a ruined condition of neglect (Key Coffee 2001). Indeed, exports of *arabica* coffee from Makassar had plummeted from a pre-war peak of 1,798 tonnes in 1936 (Paerels 1949) to only 121 tonnes in 1973 (BPS 1974). Early product marketing by Key Coffee was based around a theme of bringing a dying coffee back to life, drawing heavily on cultural images and the spiritual mystique of Toraja to create a unique, geographically informed product image. The company maintains that they began using the Toraja identity when Sulawesi coffee was unanimously referred to internationally as Kalosi, and so claims responsibility for the construction of this particular quality association.

The importance of the Toraja geographical identity to Key Coffee was quickly established when the Sulawesi-based operating company (Toarco Jaya) took its name from an acronym of Toraja Arabica Coffee. Key Coffee registered Toraja as a Japanese trade mark in 1977, followed by an Indonesian trade mark for the company logo (a traditional Torajan house) in 1979, and a US trade mark for Toarco Toraja in 2002 (Dirjen HakI 1979; Industrial

Property Database Library (IPDL) 2003; Trademark Electronic Search System (TESS) 2003). The Japanese trade mark specifically protects against the use of the Toraja name by other roasting companies in Japan, irrespective of the actual coffee origin, whereas the US trade mark includes a disclaimer to such an exclusive right. The company has been prepared to take legal action on more than one occasion in Japan to protect their exclusive right (Key Coffee 2001). Correspondingly, a recent request by a rival coffee company to register Toraja Arabica Coffee, with an accompanying map of Sulawesi, as a trade mark in Japan was rejected in 2001 (IPDL 2003).

Despite emerging relatively recently as a major coffee-consuming country, Japan is now the world's third largest importing nation after the US and Germany (ICO 2004). Furthermore, coffee imports to Japan are purchased, on average, at a 10 per cent premium above the ICO composite indicator (ICO 1998). Japan also routinely buys most of the crop of Hawaiian Kona and Jamaican Blue Mountain (Pendergrast 2001; Association of Indonesian Coffee Exporters (AEKI) 2002), indicating a willingness to absorb the world's rarest and most expensive coffees, and a strong consumer belief in the relationship between quality and the geographical embeddedness of production.

During much of the 1980s and 1990s, Japan was easily the foremost destination for Sulawesi *arabica* coffee exports. Even though by 1999 the volume of imports to the US had begun to exceed those to Japan, the latter was still the highest value importing country during 2002 (Table 14.2). The average price of imports into Japan was US$2.9 per kilogram, compared with US$1.9 per kilogram for both the European and US markets. Within the Japanese coffee-drinking culture where place names are valuable commodities in themselves, Toraja has emerged with an enviable reputation for quality and routinely demands a premium price.

Local estate owners and Makassar-based exporters in Sulawesi have been understandably frustrated by the restrictions imposed on their ability to benefit from the place-related reputation of Toraja coffee in the lucrative Japanese market. In response to the Toraja trademark held by Key Coffee, the Association of Indonesian Coffee Exporters (AEKI) has prepared applications to register the place names of nine regional coffees across Indonesia

Table 14.2 Exports of *arabica* coffee from Makassar Port, 2002

Importing region/ country	Volume (tonnes)	Total value (US$ '000)	Average price ($US/kg)
Japan	1,194	3,496	2.93
Europe	574	1,072	1.87
US	1,612	2,893	1.89
Australasia	264	552	2.10
Total	3,644	8,013	2.25

Source: Deperindag 2002a.

as AEKI trade marks, including, somewhat ambiguously, the identity Toraja Kalosi (Kurniasih 2003). Unfortunately, in its current formulation, the AEKI proposal does not offer a mechanism for verifying geographical authenticity or quality of coffee exports, and will not ultimately affect the existing Japanese trademark. Quoted from an article in a Makassar-based newspaper, the Head of the South Sulawesi Branch of AEKI complained in frustration that,

> foreign companies have no right to claim Indonesian coffee products as their own intellectual property, as those coffee names concord with their geographical location within Indonesia. Before those foreign companies registered the Toraja coffee name in America and Japan, we were already popularizing that product.
>
> (Quoted in Fajar 2002)

The Japanese trade mark and the trading implications thereof contrast strongly with other systems of geographic protection, such as those recognized multilaterally as Geographical Indications, under the TRIPS agreement (Trade Related Aspects of Intellectual Property Rights) of the Uruguay Round. Geographical Indications provide communal protection for all producers living in a particular region against the fraudulent abuse of their place name by unqualified producers and traders. In contrast the Japanese Toraja trade mark acts to restrict otherwise geographically legitimate producers from accessing a particular national market.

Geographical Indications have been widely applied by producer associations to protect the market image and authenticity of a variety of mostly gourmet agricultural products, notably wines and cheeses in Europe. Despite the historic association between geographic origin and quality in the global coffee industry, Geographic Indications remain infrequently used in the sector. The location of many coffee-growing regions in the less affluent tropical regions of the South would appear to inhibit such protection due to the substantial costs (heavily subsidized by government institutions in countries such as France) associated with registration and ongoing supervision of Geographical Indications. While Geographic Indications are now being established in the global coffee industry, with countries such as Jamaica, Guatemala and Mexico taking lead roles, industry self-regulation of the use of place names remains dominant.

Toraja to Tokyo trade networks

The estate plantation owned and operated by Key Coffee supplied 15 per cent of the company's export requirements in 2002 (Hirosan 2002) and is located in eastern Toraja at moderate altitudes between 900 and 1,300 metres. The remaining 85 per cent was sourced via community purchasing operations, which are located in the north of Toraja to coincide with access routes to higher-altitude growing villages (up to 2,000 metres).

In itself, the remoteness of the purchasing station functions to encourage the supply of only that coffee grown in its immediate vicinity. The company insistence on accepting only semi-wet parchment coffee is also a deliberate attempt to encourage only the supply of locally grown coffee.[5] The extended storage of semi-wet parchment coffee makes it susceptible to mould formation, which is readily identified during cup-testing. Strict quality control procedures during purchasing include physical inspection, a defect count and rigorous cup-testing of each batch prior to acceptance. Consequently, the risk of rejection is high for sub-standard coffee, and the inconvenience and costs to local traders of transporting the coffee back to alternative trade networks is substantial. Thus, while this purchasing system does not entirely discount the possibility of non-local coffee being offered to the company, it strongly selects for locally grown beans.

In addition to consistently offering the highest (locally available) prices, the company pays significant premiums for defect-free coffee, and for those beans exhibiting particularly desirable cup characteristics (which the company essentially associates with higher altitude production). Nowhere else in Sulawesi are price incentives for quality so immediately transferred to growers, significantly affecting on-farm processing in northern Toraja. Key Coffee has also established a priority purchasing arrangement with Sapan village (Figure 14.1), where the company operates purchasing activities at considerable expense in the village a day prior to the local market. The hinterland for this market village includes the highest-altitude coffee gardens found in Sulawesi. Local growers and not village traders are encouraged to sell directly to the company at this weekly station. Through their own estate production, and implementation of a strict purchasing policy, the company is therefore able to ensure that virtually all their Sulawesi coffee is geographically authentic Toraja coffee. Key Coffee believes in the quality attributes of coffee grown in the villages of northern Toraja and, as a result of their highly integrated operations, is generally able to ensure first pick at this coffee.

Key Coffee continues to dominate imports into Japan (Figure 14.2). The company's share, however, was substantially reduced in 2002, when a small importer-roaster imported 414 tonnes of Sulawesi coffee into Japan (Deperindag 2002a). The coffee imported by this company in 2002 was purchased (at an export price significantly below other Japanese buyers) from a single processor-exporter with a mill in the Toraja region, and labelled as Toraja Green coffee. This centrally located mill does not implement the same purchasing system and geographical control as Key Coffee and their coffee originates from the various growing districts of South Sulawesi. More important perhaps than their relative inability to trace the geographical authenticity of their Toraja coffee are the trading restrictions in Japan

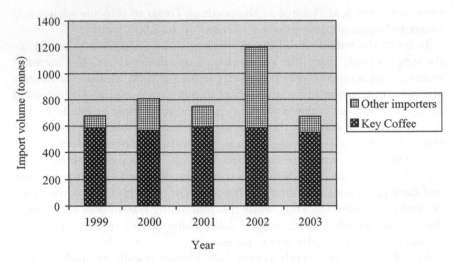

Figure 14.2 Share of Japanese import market held by Key Coffee, 1999–2003
Source: Key Coffee 2001; Deperindag 2002b.

incurred by Key Coffee's ownership of the Toraja identity. Presumably, this company was unable to benefit from the quality associations of Toraja coffee in Japan, and did not subsequently purchase Sulawesi coffee during the 2003 season.

Conclusion: regulating the benefits of geographical embeddedness

There exists a range of physical and cultural environments across sites of coffee production in Sulawesi, such that factors of geographical embeddedness affecting quality construction differ significantly. For the most part, however, quality differentiation according to actual local origins is not consistent with the subsequent geographic identities applied at the site of export. There are currently no local institutions with the political or economic will to ensure the geographical authenticity of the coffee being exported from Makassar under specific geographical identities. Despite these place names remaining principal modes of quality differentiation for importers and roasters in consuming countries, their allocation at the site of export is, to a large extent, arbitrary.

The one exception to this pattern is the vertically integrated structure of Key Coffee supplying the Japanese market. The importance of geographical specificity in the Japanese coffee market has resulted in a high degree of regulation concerning the use of geographical identities. 'Fair Competition Rules, Regulations and Guidelines concerning the Labelling of Regular Coffee and Instant Coffee' are implemented and monitored by the All Japan

Coffee Association (AJCA). These also include composition requirements for the use of geographic expressions in blended coffees (Ueshima 2001). The AJCA is also a key actor responsible for the enforcement of Key Coffee's Toraja trade mark (Kito 2002). The enrolment of the AJCA in the protection and valorization of geographical specificity as a means of constructing quality associations has been an important aspect contributing to the appreciation of geographical coffee identities in Japan.

This, then, highlights a set of contradictions with regard to regulating geographical identities in the Sulawesi coffee sector. The current structures surrounding the use of the Toraja identity along the coffee supply chain into the Japanese market are dominated by control and ownership by a non-local corporate entity. Institutional arrangements, which include legal recognition of the geographical expression as exclusive intellectual property and a supportive industry association (AJCA), are fundamental to the ability of Key Coffee to assert this control. This control is further reflected in their ability to maintain market dominance and effectively create scarcity for a particular geographical product through a legally enforced monopoly. Vertical integration of the entire supply chain (to plantation) and tight control of community purchasing operations in northern Toraja ensure that any economic benefits of the geographical associations of quality are retained within the company.

While coffee producers and exporters in Sulawesi reasonably object to the appropriation of a local geographical expression as foreign property, the case of Toraja coffee is complicated by the instrumental role performed by Key Coffee in the initial construction of geographical quality associations within the Japanese market. The company's exclusive rights in Japan are also supported by their Sulawesi-based operations, which take great efforts to enforce geographical authenticity during community purchasing. These factors are emphasized by the company as the *raison d'être* for their exclusive trade mark.

The progressive dismantling of state support for the domestic coffee sector in many producing countries, including Indonesia, corresponds with increasing penetration of international coffee companies and new forms of industry regulation. In the specialty coffee sector, the use of geographical identities as a means for quality differentiation is an important component resulting in corresponding changes in supply chain governance. There are, it would seem, potential economic benefits for producers able to control the quality associations of their geographic embeddedness. However, the case of Toraja coffee in the Japanese market demonstrates that these benefits are not self evident in the popularity and consumer recognition of these associations. Corporate self-regulation of geographical authenticity in other non-Japanese consuming markets is equally problematic, as depicted by the misuse of the Sumatran Mandheling identity for Sulawesi exports to the US, Singapore and Europe. The level of public support required for the collective protection of place names as a Geographical Indication may be untenable in many producing country contexts. Moreover, there continues to be ongoing debate in the WTO about the status of such institutional support as constituting an illegitimate agricultural subsidy.

The increased consumer interest in regional food products suggests that any attempt to understand the multi-faceted reality of agri-food globalization should incorporate the structural trajectories implied by geographic specificity on supply chain structures and trade relationships. This case emphasizes the role of institutional arrangements in determining both the nature of supply chain relationships and the distribution of economic benefits among supply chain actors. Within the coffee supply chain extending from Toraja through to Tokyo, the regulation of geographical specificity has emerged as an important mode of economic ordering. Unlike the re-regulation of agri-food trading systems elsewhere resulting from producer initiatives to protect geographic identities, consumer and near-consumer demands for geographical authenticity are orchestrating the re-regulation evident here. The particular institutional arrangements regulating geographical embeddedness and the use of place names clearly affect the potential for quality-related upgrading by rural producers. Without appropriate institutional support, any values associated with the geographical embeddedness of agricultural production can be effectively controlled and regulated by corporate interests rather than local producers.

Notes

1 The Toraja region is also known by its official name, Tana Toraja, which refers specifically to the administrative division. For the purposes of this chapter, the two terms are considered synonymous.

2 The export data for Sulawesi (including volume, prices, destinations, international buyers and geographical identities) presented in this chapter was compiled from original export documentation (ICO Certificates of Origin and Export Notification Certificates) sighted by the author at the provincial office of Industry and Trade in Makassar.

3 A highly publicized case of geographical fraud in the coffee industry was the 'Kona Kai Scandal', which apparently 'sent shock waves through the specialty sector' (Pendergrast 2001: 315). In 1996, it was exposed that a California importer was buying cheap Panamanian and Costa Rican beans and rebagging them as Kona. From the evidence in Sulawesi it appears that such practices are more widespread than acknowledged by many industry actors, highlighting the current difficulties in ensuring geographical authenticity in the industry.

4 The Toraja highlands are the foremost tourist attraction in eastern Indonesia due to a unique and complex traditional belief system, ornately carved architecture and a vibrant ceremonial cycle. These characteristics have more recently become a marketing resource in the specialty coffee sector.

5 In northern Toraja, coffee cherries are pulped, fermented and dried for only a few hours by the farmer before being sold to local market traders. In this semi-wet condition, the coffee is purchased by hulling operations. Key Coffee then mechanically dries the coffee prior to hulling, while it is common for other hullers to hull the semi-wet parchment prior to sun-drying the green beans.

References

Association of Indonesian Coffee Exporters (AEKI) (2002) 'Jamaica's Blue Mountain Coffee: the most expensive in the world', *Kopi Indonesia*, 108 (original publication: *Tea & Coffee Trade Journal*, September 2001): 16–18.

Badan Pusat Statistik (BPS) (1974) *Ekspor Menurut Jenis Barang, Negeri Tujuan dan Pelabuhan Ekspor*. Jakarta: Badan Pusat Statistik (National Statistics Agency).

Badan Pusat Statistik (BPS) (2002) *Kontribusi PDRB di Tana Toraja*. Rantepao: Badan Pusat Statisk Kabupaten Tana Toraja (Statistics Agency).

Barham, E. (2003) 'Translating terroir: the global challenge of French AOC labeling', *Journal Of Rural Studies*, 19 (1): 127–38.

Bigalke, T. (1981) *A Social History of Tana Toraja. 1870–1965*. Madison (WS): PhD dissertation, History Department, University of Wisconsin.

Deperindag (2002a) *Pemberitahuan Ekspor Barang (PEB) Export Certificates (raw data)*. Makassar: Department of Industry and Trade.

Deperindag (2002b) *Realisasi Ekspor Kopi Sulawesi Selatan tahun 1999 S/D 2001*. Makassar: Department of Industry and Trade, Provincial Office of South Sulawesi.

Dirjen HakI (1979) 'Indonesian Trademark No. 141918 held by PT Toarco Jaya', Jakarta: Directorate General of Intellectual Property Rights, Department of Justice.

Fajar (2002) 'AEKI Daftar Kopi Toraja ke Departemen Kehakiman' [AEKI moves to register Toraja coffee with the Department of Justice], Makassar: Fajar, 5 May (translated by J. Neilson).

Granovetter, M. (1985) 'Economic action and social structure: the problem of embeddedness', *American Journal of Sociology* 91 (3): 481–510.

Hirosan [pseudonym] (2002) Personal communication. Purchasing Manager, PT Toarco Jaya. Rantepao, 16 August.

International Coffee Organization (ICO) (1998) *Importing Members – Japan Country Profile*. London: International Coffee Organisation. Online: www.ico.org.au (accessed 27 January 2004).

International Coffee Organization (ICO) (2004) *Historical Data – Imports of Importing Countries*. London: International Coffee Organisation. Online: www.ico.org.au (accessed 27 January 2004).

International Coffee Organization (ICO), International Trade Centre (ITC) and Common Fund for Commodities (CFC) (2000) *The Gourmet Coffee Project: Adding Value to Green Coffee*. London: ICO, ITC, CFC (UN Conference on Trade and Development (UNCTAD)).

Ilbery, B. and M. Kneafsey (1999) 'Niche markets and regional specificity food products in Europe: towards a research agenda', *Environment and Planning A*, 31: 2207–22.

Industrial Property Digital Library (IPDL) (2003) *Industrial Property Digital Library*. Online: www.ipdl.jpo.go.jp (accessed 19 August 2003).

International Trade Centre (ITC) (2002) 'Chapter three: niche markets, environment and social aspects. Coffee: an exporter's guide', Geneva: ITC/UNCTAD/WTO.

Key Coffee (2001) *Perjalanan Panjang Usaha Kopi di Tana Toraja* [*The Long Journey of a Coffee Business in Tana Toraja*]. Rantepao: Key Coffee Inc (translated by Jeffrey Neilson).

Kito [pseudonym] (2002) Personal communication. Director of Production, PT Toarco Jaya. Rantepao, 19 August.

Kurniasih, T.J. (2003) Personal communication. Intellectual Property Development, AEKI. Jakarta, 4 September.

Lewin, B., Giovannucci, D. and Varangis, P. (2004) *Coffee Markets: New Paradigms in Global Supply & Demand*. Washington (DC): World Bank – Agriculture and Rural Development.

Moran, W. (1993) 'Rural space as intellectual property', *Political Geography*, 12 (3): 263–77.

Paerels, B. (1949) 'Bevolkingskoffiecultuur'. In J.J.V. Hall and C.V.D. Koppel (eds) *De Landbouw in de Indische Archipel*. The Hague: N.V. Uitgeverij W. Van Hoeve: 89–119.

Parrot, N., Wilson, N. and Murdoch, J. (2002) 'Spatializing quality: regional protection and the alternative geography of food', *European Urban and Regional Studies*, 9 (3): 241–61.

Pendergrast, M. (2001) *Uncommon Grounds: The History of Coffee and How it Transformed the World*. New York: Texere.

Ponte, S. (2002a) 'The "latte revolution"? Regulation, markets and consumption in the global coffee chain', *World Development*, 30 (7): 1099–122.

Ponte, S. (2002b) 'Standards, trade and equity: lessons from the specialty coffee industry', *CDR Working Paper*, 02.13. Copenhagen: Centre for Development Research.

Ray, C. (2001) 'Transnational co-operation between rural areas: elements of a political economy of EU Rural Development', *Sociologia Ruralis*, 41 (3): 279–95.

Specialty Coffee Association of America (SCAA) (1999) *Coffee Market Summary*. Long Beach (CA): Specialty Coffee Association of America.

Specialty Coffee Association of America (SCAA) (2004) 'What is specialty coffee?'. Online: www.scaa.org/What_is_specialty_coffee.asp (accessed 5 December 2004).

Trademark Electronic Search System [TESS] (2003) US Patent and Trademarks Office. Online: www.uspto.gov (accessed 19 August 2003).

Ueshima, T. (2001) 'Overview of the coffee market in Japan', *First World Coffee Conference*, 17–19 May. Online: www.ico.org/event/wcc.htm (accessed 27 January 2004).

Ukers, W.H. (1935) *All About Coffee*. New York: The Tea & Coffee Trade Journal Company.

15 Globalization, the WTO and the Australia–Philippines 'banana war'*

Robert Fagan

Introduction

In August 2003, the government of the Philippines took Australia's long-standing ban on banana imports to the World Trade Organization (WTO). The ban is aimed at preventing a range of banana diseases and pests entering Australia that could devastate its small but significant banana industry, which supplies the domestic market but does not export. The Philippines argued that this quarantine measure was not justified on scientific grounds and was illegal under WTO phytosanitary (plant health) rules. From Australia's perspective, this challenge presented a potentially important precedent with regard to the future of the nation's quarantine system, with implications that would flow well beyond the banana sector. More broadly, this case provides an illustration of the *multi-scalar* forces that interact in the construction of cross-continental food chains and, more particularly, the international politics of agri-food trade.

This chapter explores the context and detail of this dispute to illuminate the political framework and economic structures of both banana commodity chains and the attempts by the WTO to enforce multilateral trade liberalization outcomes. It examines: first, conceptual frameworks for examining international agri-food relations; second, the national politics of recent quarantine policy affecting the Australian banana industry; third, 'global' threats to the industry identified by banana growers through their peak organization (the Australian Banana Growers' Council (ABGC)); and, fourth, the way in which the recent dispute with the Philippines is being constructed simultaneously at several geographical scales. The chapter concludes with a brief review of the role of transnational corporations (TNCs) in the dispute and an assessment of WTO regulation of this issue.

* The research for this chapter was funded through Australian Research Council Discovery Grant 'The Spatial Construction of Food Commodity Chains'. The author is grateful for the assistance of Leah Gibbs and Hélène Mountford in tracking down information about the global banana industry and local responses to the quarantine inquiries; to his fellow grant-holders Bill Pritchard and David Burch for useful discussions about the conceptual framework for this research; and to Jasper Goss of the International Union of Foodworkers (Asia and Pacific Regional Secretariat) for useful comments and information.

Table 15.1 Exports of Cavendish dessert bananas (major exporters, 1995–2000)

Exporter	Exports (kilotonnes)						
	1995	1996	1997	1998	1999	2000	% of total 2000
Ecuador	3,737	3,840	4,456	3,848	3,865	3,932	37.2
Costa Rica	2,033	1,933	1,835	2,101	2,113	1,814	17.2
Columbia	1,336	1,407	1,500	1,436	1,650	1,506	14.3
Philippines	1,213	1,271	1,154	1,150	1,320	1,418	13.4
Panama	693	634	602	463	593	538	5.1
Guatemala	636	611	659	794	680	527	5.0
Honduras	522	574	558	502	109	150	1.4
Others (inc. ACP)	1,264	1,369	1,383	1,377	1,388	675	6.4
Total	11,434	11,639	12,147	11,671	11,718	10,560	100.0

Source: Biosecurity Australia 2002: 34.

Prior to these discussions, however, some background to the global banana industry is required. Table 15.1 and Figure 15.1 describe the global composition of Cavendish banana production, the dominant variety grown for international trade. This geographical pattern reflects a complex mixture of: changing fortunes among the TNCs that dominate the banana trade; local production conditions in South and Central America; preferential access regulations granted by the European Union (EU) to former colonies in Africa, the Caribbean and the Pacific ('the ACP countries'); and the connections between American domestic politics and positions taken by the US government in the WTO.

The political and economic complexities underpinning global, national and local issues in the banana sector were laid bare by the acrimonious 'banana war' between the US and the EU during the 1990s. Significantly, this was the first major dispute brought to the WTO after its establishment in 1995. The world's big three banana-trading TNCs, Chiquita Brands, Dole and Del Monte, are all either US-owned or represent substantial American interests. By 1997 these TNCs controlled two-thirds of world trade in bananas (van de Kasteele 1998) and supplied developed country markets, including those in the EU, from a mixture of plantations and contracted growers in the tropical producing countries of Table 15.1. The EU's common banana import regime introduced in 1993, however, guaranteed shares of its market to the ACP countries, despite their higher production costs compared with the major South and Central American producers.

In 1995 the US government, on behalf of its banana TNCs, filed a WTO complaint against this EU policy and, subsequently, the WTO found against the EU. Nevertheless, conscious of severe economic impacts on ACP producers if preferences were abandoned (McMahon 1998: 104), the EU simply

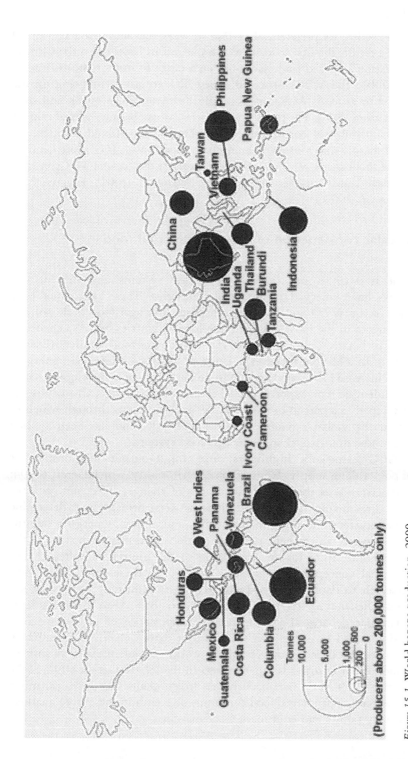

Figure 15.1 World banana production, 2000

Source: Adapted from FAO 2001.

refused to modify its preferential system in ways that addressed the WTO ruling. As a result, the first 'banana war' broke out in 1999 when the Clinton Administration imposed trade sanctions on a range of imports from Europe. In April 2001, the WTO brokered a new EU arrangement bringing the banana war to an end. ACP quotas would be replaced by tariffs and then phased out from 2006. Yet the peace has been uneasy because global market shares remain political constructions and the deal negotiated through the WTO had sharply different impacts on the banana TNCs. (For more detailed accounts of this banana war, and the central role of the banana TNCs in relation to American domestic politics, see Brimeyer 2001; Fagan 2002; Hermann *et al.* 2003.)

Multi-scalar construction of cross-continental food chains

Since the mid-1990s new global governance systems and institutional frameworks have emerged in agri-food industries. These reflect not only formation of the WTO and continued disputes over agri-food trade policies of governments in the most powerful developed countries but also changing relationships between agri-food TNCs and corporate food retailers (Barrett *et al.* 1999). In addition, however, a range of agri-food NGOs has emerged, seeking to increase local social and environmental justice in developing countries dependent on food exports. Global commodity chain (GCC) analysis has been developed as a partial but very useful 'analytical lens through which to understand the global economy' (Gereffi 1994: 96) and has been applied widely to these dramatic changes in agri-food systems.

The GCC framework highlights: first, input–output relationships at various points along supply chains; second, territoriality, which Gereffi (1994: 96–7) understands as the spatial dimensions of change; third, new governance structures involving power relations between firms, which determine resource allocations and flows in the chains; and, fourth, emerging institutional frameworks which shape processes of globalization at each stage in the chain (see Ponte 2002 for an application to the globalized coffee chain). Despite equal conceptual significance attached to these four dimensions, GCC research to date has been dominated overwhelmingly by governance structures (Dicken *et al.* 2001: 99). In recent agri-food research, this has reinforced familiar debates about the relative importance of producer-driven chains, dominated by large TNCs, and buyer-driven chains dominated by supermarkets and brand managers in the world's major markets. A shift in the balance of power towards supermarkets in the 1990s (Gereffi 1994; Leslie and Riemer 1999: 403) has made buyer-driven chains pervasive, so much recent research has privileged food consumption trends in these 'core' markets as both entry points and culminations of analysis.

In exploring the Australia–Philippines banana dispute, this chapter focuses on recent changes outside the world's core markets. In particular, it focuses on aspects of territoriality and institutional frameworks, which remain less developed overall in GCC research. The focus on global governance structures, and transnational frameworks such as the WTO, has tended to sideline the different institutional configurations characteristic of different nation-states in GCC analysis (Dicken *et al.* 2001: 100) despite these remaining crucial in agri-food systems. Neither dominant neo-liberal accounts, equating globalization with trade liberalization, nor 'hyperglobalist' accounts (Held *et al.* 1999: 3–5), still fixated on the power of transnational corporations (TNCs), provide adequate understandings of the *multi-scalar* forces involved in either shaping institutional frameworks or producing new territorialities in commodity chains such as bananas.

The importance of specific place-based practices of production and consumption (Hughes 2000), and their fluid relationships with national scale institutional frameworks, are easily obscured if it is simply assumed that scales such as global, national and local exist in a pre-given hierarchy dominated increasingly by global forces such as TNCs or the supermarket-driven supply chains. This chapter explores how the dispute between the governments of Australia and the Philippines over bananas has been constructed politically (see Howitt 2003) at global, national and local scales simultaneously.

The Australia–Philippines 'banana war'

Australia is unique among developed countries in being self-sufficient in bananas. The overwhelming majority of Australia's bananas are grown in tropical north Queensland, with significant production also in sub-tropical regions of south-east Queensland and northern New South Wales (NSW) (Table 15.2, Figure 15.2). For comparison, Australian total production in 2000 was less than half of that in Guatemala or Panama, two of the smaller Central American producers. NSW was the largest producing state until 1980 after which it was overtaken by Queensland, following rapid expansion in the Tully-Innisfail region south of Cairns. Banana farming is carried out at larger scale in the tropical localities, with yields of 34 tonnes per hectare achieved in north Queensland in 2000, broadly comparable with yields achieved in Mindanao, the major production region in the Philippines. This compares with yields of only 18 tonnes per hectare reported for northern NSW (Biosecurity Australia 2002: 36). By 1993 the 26 largest banana farms in north Queensland produced 21 per cent of total Australian banana production (Borrell *et al.* 1993: 7). Valued at $A350 million (approximately US$200 million) in 2002 and involving about 2,000 farmers (Biosecurity Australia 2002: 35), banana-growing is a modest industry within Australian agriculture, but nevertheless one with significant regional impacts.

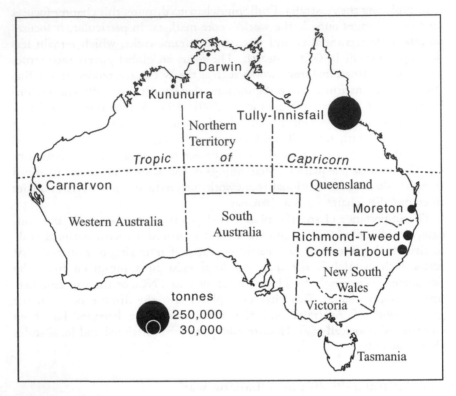

Figure 15.2 Banana-producing districts in Australia, 2003
Source: Prepared from data in ABGC 2004b.

Initially, the high costs of transporting bananas from Central and South America afforded geographical protection to Australia's relatively high-cost industry. From the early 1970s, however, falling relative costs of shipping amplified steadily falling world prices of bananas, and low-cost bananas became potentially available from the Philippines. Yet Australian banana-growers remained protected by tight phytosanitary quarantine regulations justified to keep its biosphere free from the wide variety of diseases and banana-loving insect pests flourishing in most tropical-producing countries, including the Philippines. These restrictions have constructed a domestic banana market in which Australian consumers are serviced only by higher-cost domestic producers. Writing in the early 1990s, Borrell *et al.* (1993: 10) estimated that in the absence of quarantine restrictions, Ecuadorian bananas could have been landed in Australia at prices 16 per cent below those prevailing at Sydney's major wholesale fruit and vegetable market (Flemington). In the late 1990s, it was observed that Australian retail banana prices were on average about twice those in the US and New Zealand (James and Anderson 1998: 434).

Table 15.2 Banana growers and production in Australia, 2001

Region	Growers		Production	
	Number	Percentage of total	Tonnes	Percentage of total
Carnarvon	74	3.7		
Kununurra	10	0.5		
Western Australia	84	4.2	8,606	3.6
Northern Territory	4	0.2	3,575	1.5
Far north Queensland	591	29.7		
South-east Queensland	366	18.4		
Queensland	957	48.1	206,869	86.0
Far north coast	435	21.8		
Mid-north coast	512	25.7		
New South Wales	947	47.5	21,359	8.9
Australia	1,992	100.0	240,409	100.0

Source: ABGC 2002d: 8.

Ecuador challenged Australia's banana quarantine regime in 1991, but risk analysis by the Australian Quarantine Inspection Service recommended the continuation of bans. After the formation of the WTO in 1995, the Philippines government sought access to the Australian market for mangoes, also protected by quarantine, and extended this to Cavendish bananas in 1999. This triggered another inquiry by the quarantine service, newly named as Biosecurity Australia (Biosecurity Australia 2002: 29). Plant and animal quarantine issues are covered by the WTO Sanitary and Phytosanitary (SPS) agreement, which recognizes the right of countries to protect human, animal and plant life and health from pests and diseases. Yet regulation must be based on scientific analysis and 'not be maintained without sufficient evidence' (Biosecurity Australia 2002: 20). In addition, quarantine regulations must not be more trade restrictive than necessary to achieve the level of protection justified by the science (James and Anderson 1998: 425).

Working within these legal obligations, in July 2002 Biosecurity Australia released for public comment draft recommendations from its inquiry into banana quarantine measures (Biosecurity Australia 2002). The key recommendation from its report was that the Australian government should continue to ban banana imports from the Philippines. A range of risks were identified, in particular the plant disease *moko* (a tree killing bacterium which cannot be recognized easily in the fruit itself) widespread in Mindanao. *Moko* is absent from Australia, and Biosecurity Australia argued the precautionary principle should apply, thus justifying maintenance of the import restriction.

This recommendation was hailed by the ABGC as consistent with the SPS agreement (ABGC 2002b), and received support from major Australian political parties. Not surprisingly, there was a different response from the

Philippines government, which challenged the scientific legitimacy of the recommendation, and announced that it might take action against Australia at the WTO. Following unsuccessful bipartite discussions that sought to resolve the Philippines' concerns, this action was eventually initiated. In August 2003 the Philippines government made a formal request to the WTO to establish a panel under the SPS agreements to adjudicate on Australia's quarantine policy, claiming that the scientific evidence did not justify contin-uation of the import bans.

Prior to the WTO Panel making a determination on this issue, however, Biosecurity Australia startled the Australian banana industry by releasing a 'revised draft import risk analysis' in February 2004 based on its considera-tion of responses by various stakeholders to its 2002 report (Biosecurity Australia 2004). Crucially, Biosecurity Australia concluded that so long as certain conditions were met (including identification of plantations in Mindanao with acceptably low levels of *moko* and other diseases, and chem-ical and other treatments of bananas during packing in the Philippines), 'the . . . risk [of importing bananas from the Philippines] . . . was in fact acceptable' (Biosecurity Australia 2004: 6). Nevertheless, the Philippines government maintains the need for more generous and permanent access and, in search of an outcome to this issue that is binding under international trade law, has continued to press its case against these measures at the WTO. At the time of writing, the ultimate outcome from this litigation has not been determined.

Seen through neo-liberal eyes, the narrative detailed above could be presented as evidence of a rational, scientific, rules-based regime operating to resolve international agri-food trade disputes. Yet as revealed in the following section, such interpretation is superficial in terms of how and under what circumstances international agri-food trade relations are constructed. The legal process of WTO negotiation is, in fact, carried out against the background of specific mixtures of local, regional and national interests, with significant involvement at local and national scales by stakeholders also operating at global scale. Hence, trade outcomes need to be understood as being orchestrated through political contestations occurring simultaneously at a range of scales. Exploring these circumstances is vital not only in under-standing the reality of this dispute but also in forging a more complete conceptual framework for how economic processes represent scaled and inter-active *political* constructions.

The politics of scale in the banana dispute

The threat of Filipino (and other) bananas entering the Australian market galvanized the local industry in several ways which, when taken together, have amounted to constructing and prioritizing a *national* frame of reference. From the local industry's perspective, quarantine restrictions provided economic protection to local producers in a situation of global over-supply and low prices, and phytosanitary protection against imported pests and diseases. Both these threats were articulated by the ABGC as being 'a consequence of globalisation' (ABGC 2002c: 13). Consequently, the industry's efforts to maintain quarantine restrictions have been subsumed into a wider critique of the Australian government's stance on globalization and trade liberalization.

First, the ABGC responded stridently to the request for market access by the Philippines government. The Council hired an experienced legal team to represent their interests in the inquiry and, in addition, instituted a levy on banana growers of 10 cents (approximately US$0.07) per 13-kilogram carton to establish a fighting fund against imports. Strong imagery was employed in this campaign to help construct banana imports as a national issue with bananas described as an 'iconic' Australian fruit. 'Aussie banana' giveaways were instituted at Brisbane markets, and miniature Australian flags were pinned to bananas (ABGC 2002a).

Second, the ABGC responded to these threats by orchestrating national-scale priorities for the industry's future, collected under the umbrella of a strategic plan released in 2002 (ABGC 2002c). The plan identified four high-priority strategies to help secure the future of the Australian banana industry: first, developing niche market exports such as organics to the Asia-Pacific Region; second, achieving a substantial increase in per capita domestic consumption levels (currently 14.5 kilograms annually compared with 20 kilograms in New Zealand); third, growing more non-Cavendish varieties, such as Lady Finger bananas in northern NSW; and fourth, increasing supply chain efficiencies (ABGC 2002c). This last strategy reflects the increasing market power in Australia's fresh horticulture industry being exerted by large supermarkets, a trait in common with experiences elsewhere across the world (Burch and Goss 1999). In recent years supermarkets have rapidly rationalized fruit and vegetable supply bases, encouraging fewer and larger wholesalers (Bunt 2002: 434). In 2003, Australia's two leading supermarket chains accounted for approximately 70 per cent of national retail sales of bananas. Indeed, the geographical shift of Australian banana production northwards to tropical Queensland was closely connected to demands by the supermarket chains for greater volume shipments of larger (and more yellow) product (Borrell *et al.* 1993: 8).

The banana industry's claims of the economic and phytosanitary dangers that would accompany relaxation of quarantine measures were also brought into sharp focus by two local events that interrupted domestic supply arrangements. In April 2001, banana plants with black sigatoka disease were discovered on a farm in the Tully district. Responses were immediate. Movement of bananas from north Queensland to NSW markets was banned through an agreement between the NSW and Queensland governments, and an eradication programme financed jointly by the Australian and Queensland governments was declared successful by June 2002. Further complicating domestic supply arrangements, in April 2002 banana workers in north

Queensland went on strike in protest against rumours (unfounded) of an in-principle agreement by the Australian government to permit some imports of bananas and pineapples from the Philippines (Australian Broadcasting Corporation 2002).

These events conditioned the progress of the dispute within Australia. While there is no suggestion of impropriety in the scientific deliberations of Biosecurity Australia, the lobbying efforts of the local industry evidently enabled considerable marshalling of arguments in favour of restrictions being retained. Moreover, political sensitivities on the future of this industry helped shape the broad tenor of submissions to the inquiry from various state and federal government agencies. Hence, national-scale resolution of this issue was constructed by local-scale political contests and interactions.

The construction of this issue in the Philippines was also conditioned by a distinctive set of regional-scale concerns. A volatile combination of local, national and, increasingly, global issues has characterized that country's export banana industry. Bananas are the country's second-largest agri-food export (after coconut products). In the late 1960s, the Marcos government developed the Cavendish dessert banana export industry both as a foreign-exchange earner and regional development vehicle for Mindanao, one of the country's poorest regions with a large Islamic population. Seeking to supply Japan, the TNCs Del Monte, Dole and Chiquita established plantations, often in partnership with leading families from the Filipino élite (Krinks 2002). Contracts were offered to local farmers to switch into banana produc-tion, with major social impacts (see Chapter 9). The Marcos government also used military personnel to enforce compulsory land acquisition, and used both the police and military to maintain order and prevent plantation workers from forming unions (Krinks 2002). By the 1980s, the plantations were incurring extra costs for security against sabotage by Muslim separatists but rates of unionization of banana workers have remained low. By 2001, national production of bananas of all varieties in the Philippines was 4 million tonnes, about half of which was Cavendish bananas, nearly all exported (Biosecurity Australia 2002: 34).

The Filipino banana industry grew largely on the back of increased import demand from Japan. The Japanese market for bananas expanded rapidly during the 1970s and 1980s, and low-cost bananas from the Philippines replaced historical supply sources in Taiwan. Since the early 1990s, however, the Japanese market for bananas has shown signs of saturation. While bananas remain Japan's largest single fresh produce import, markets for other fresh fruits and vegetables have grown more rapidly. By 2000, Filipino bananas still held 75 per cent of the Japanese market, but their market share was under increasing threat from Ecuador and China.

For the Philippines, the danger of losing market share in Japan posed considerable regional political implications. In the production island of Mindanao there has been a long-running separatist conflict between the Philippine government and Muslim guerrillas. In the global political climate following 11 September 2001, this conflict has taken new significance and meaning. Thus, when Australian Prime Minister Howard met with President Arroyo of the Philippines in July 2003, ostensibly to discuss progress in the Asia-Pacific component of the 'war on terror', President Arroyo linked long-running Muslim terrorist activities in Mindanao with local poverty. Regional development, she argued, was clearly linked to the fortunes of agri-food export industries (although see Lockie, this volume). On these grounds, she sought Australia's cooperation in opening its domestic market to Philippine fruit, especially bananas.[1]

At the same time, moreover, the banana dispute spilled over into the wider trade politics of Australia and the Philippines. Representatives of the Philippines government and trade associations indicated that Australia's position with regard to bananas could elicit retaliatory responses that would harm Australian agri-food exports to the Philippines (which include beef, fresh vegetables and, especially, dairy products such as powered milk and cheese: O'Loughlin 2002). This could take place through decisions by the Philippines government to reallocate import quotas away from Australia or, in the case of the WTO deciding in favour of the Philippines, for specific tariffs to be levied on Australian imports. The Philippines–Australia Business Council in Manila and representatives of the Australian dairy industry pointed out that Australia's annual banana industry turnover is lower than the current value of its dairy exports to the Philippines (O'Loughlin 2002; Bonlac Foods *et al.* 2003). The Australian government, they said, had a responsibility to protect these exports coming from a deregulated industry which had been something of a success story under the Howard government's 'Supermarket to Asia' strategy for fresh and processed foods (Pritchard 1999). As is so often the case in such agri-food disputes, local interests in one set of agricultural regions (north and south-east Queensland, northern NSW) were pitted against those of another (Victorian milk-producing districts).

In these multi-scaled politics of trade, reactions in Australia were inevitable. A spokesman for the ABGC said: 'we want this process to be based on science only. This is not about trade and certainly not about terrorism; it's about disease security that protects our island nation from new pests and diseases' (*Innisfail Advocate* 2003: 3). Echoing these local and regional concerns, the Australian Federal Opposition argued that a bilateral decision by the government favouring the Philippines request, done in the spirit of actively seeking cooperation in the war against terrorism, would undermine the scientific basis of SPS agreements and set a precedent for more concerted campaigns to undermine Australia's quarantine protection. This issue did not go away in succeeding months. In February 2004, the ABGC argued: 'if the Federal Government does an about-face and allows imports, we can only assume that this is a political decision rather than one based on science' (ABGC 2004a). In response to Biosecurity Australia's reversal of its earlier recommendation, Prime Minister Howard said: 'we are very proud of the scientific base of our quarantine approach and we do not intend to depart

from that. My information is that some new evidence was presented'
(Australia: Hansard 2004a: 25299). The Federal Minister for Agriculture
denied allegations of political pressure being applied to Biosecurity Australia
associated with the WTO reference, or bilateral negotiations with the
Philippines government over Australia's beef and dairy exports (Higgins
2004). Yet to mollify local interests, he announced commissioning of an
economic impact assessment of banana imports on the Australian industry.

Conclusions

The dispute over bananas between Australia and the Philippines is being
constructed simultaneously at several geographical scales. A multi-scalar
approach strengthens understanding of both new spatial configurations and
the emergence of new institutional frameworks which will shape the future
of the banana industry in Australia. Focusing on the national scale, James
and Anderson (1998) argue that through lifting its quarantine restrictions
Australia would gain by becoming less vulnerable to repeated challenges at
the WTO. This general position carries considerable weight with the Howard
government, anxious to retain its credibility as an advocate of multilateral
agricultural trade liberalization. The government is also sympathetic to
arguments of agricultural economists that allowing imports would increase
the welfare of domestic consumers while meeting justified claims by developing countries to sell products like bananas to rich countries. To others in
Australia, especially in banana-growing regions and localities, a decision to
allow imports would signal increased competition for a locally significant
industry from low-cost bananas produced in countries where environmental
and labour standards are low, and generate phytosanitary threats to disease-
free industries. Such threats necessitate what the regionally based Banana
Growers' Council recently described as 'adequate border protection'. But
regional and local interests are not represented simply by banana farmers and
the communities in which they are embedded. They also include powerful
national political interests, regionally based through federal electoral contests
in marginal seats, local operations of national wholesaling and retailing businesses, local farming interests in other tropical commodities such as sugar
facing uncertain export markets, and dairy farmers in eastern Victoria.

Global commodity chain analysis provides a useful framework for understanding the events explored in this chapter, but its focus on structures of
governance, especially the role of TNCs in relation to the growing power
of food retailers, needs to be extended by research that explores institutional frameworks and multi-scalar interactions between stakeholders at all
points along the chain. 'Actually existing globalizations'[2] remain highly
uneven in agri-food industries, even apparently simple ones such as those
delivering Cavendish dessert bananas through supply chains involving a
handful of powerful TNCs to the world's affluent food markets. In particular, this chapter has highlighted ways in which the 'banana war' between

Australia and the Philippines has been scaled and rescaled by actors participating in the drama and constructing stories about globalization and its local
impacts.

While the banana TNCs played a central role in the banana war between
the world's two largest markets, the United States and the European Union,
their role in the Australia–Philippines dispute has been markedly different.
Corporations such as Del Monte and Dole with a major presence in the
Philippines would favour opening the Australian market to imports, but
the example of New Zealand suggests bananas supplied from Ecuador by
Dole and Ecuadorian company Noboa would be highly competitive in
Australia if quarantine regulations were relaxed across the board. The 'big
three' TNCs are likely to remain more interested in the growing markets
of China and newly industrializing countries of the Asia-Pacific Region for
bananas sourced from their Filipino operations. For an interlocking set of
national and regional political reasons, the government of the Philippines
and 'local' producers – dominated as they are by members of the national
political élite – are more centrally involved in driving the dispute with
Australia.

Within the Australian banana industry, the role of the TNCs has been
limited and recent, underscoring the different territorial and institutional
framework for this Antipodean banana war. In 1990, Chiquita became the
only one of the big three to involve itself directly in Australian production,
purchasing some large farms in the Tully-Innisfail area of north Queensland.
Yet, despite initial fears of banana growers in Tully-Innisfail, Chiquita
Brands South Pacific Ltd (CBSP) signalled after 1997 that its Australian
intentions were about diversification and leveraging its Chiquita brand
name rather than producing bananas. CBSP acquired significant shares of
Australian blueberry and mushroom production (1997–2000), effectively
sold out of banana growing in 2002 (Chiquita Brands South Pacific Ltd,
2003) and merged its wholesaling interests with one of Melbourne's principal
fruit and vegetable wholesalers to develop direct supply chains with Coles
and Woolworths supermarkets. These local corporate reconstructions have
sharply reduced Chiquita International's equity in CBSP but the TNC retains
close affiliation through use of brand names and marketing connections. By
2000, CBSP had secured at least 20 per cent of wholesale trade in Australian
bananas through these relationships with the supermarkets and now appears
to be an effective 'stalking horse' for distribution of imported bananas from
Chiquita's global operations and those of its global competitors.

The example of the Australian banana industry shows the paramount
importance of national and local issues facing attempts by the WTO to
enforce multilateral trade liberalization regimes (see Brimeyer 2001) and
which also shape potential impacts of WTO decisions within specific national
institutional frameworks. It seems unlikely that much of Australia's sub-
tropical banana industry could survive significant relaxation of its phytosanitary quarantine regulations except perhaps by focusing on special varieties

or organic farming. Consumers may be prepared to pay a premium to support specific localized production against 'the global banana'. Larger farmers in north Queensland would fare better while, according to estimates made by Borrell *et al.* (1993), small producers even in the tropical north would have trouble being competitive with import parity prices especially in years of glut on world markets. This poses a significant dilemma for the Australian government given the interests involved in local agriculture, both in banana regions and other food sectors either affected by trade disputes or protected by similar quarantine restrictions.

Notes

1   In passing, it is worth noting that the negotiations about bananas cannot be separated from their specific national context in Filipino politics. The current Minister for Agriculture, Luis Lorenzo, is chairman of Lapanday (Australia: Hansard, 2004b: 20285) one of Mindanao's largest banana operations and affiliated to Del Monte.
2   With apologies to Brenner and Theodore (2002).

References

Australia: Hansard (2004a), *House of Representatives*, Thursday 19 February: 25299.
Australia: Hansard (2004b), *Senate*, Monday, 1 March: 20285.
Australian Banana Growers' Council (ABGC) (2002a) 'On Australia Day be aware that an Aussie icon is at risk', *Media Release*, 24 January. Online: www.abgc.org.au (accessed 27 September 2002).
Australian Banana Growers' Council (ABGC) (2002b) 'Banana growers welcome objective decision-making', *Media Release*, 2 April. Online: www.abgc.org.au (27 September 2002).
Australian Banana Growers' Council (ABGC) (2002c) *Australian Banana Industry: Strategic Plan*. Rocklea (Qld): ABGC Council.
Australian Banana Growers' Council (ABGC) (2003b) 'Regional peace issues cloud banana import science', *Media Release*, 15 July. Online: www.abgc.org.au/pages/media/030718_094440.asp (accessed 30 July 2003).
Australian Banana Growers' Council (ABGC) (2004a) 'Industry expects Howard to hold the line against banana imports', *Media Release*, 16 February. Online: www.abgc.org.au/pages/media/040216_151404.asp (accessed 5 April 2004).
Australian Banana Growers' Council (2004b) 'The industry', *Industry Statistics*, 6 July. Online: www.abgc.org.au/pages/industry/bananaIndustry.asp (accessed 6 July 2004).
Australian Broadcasting Corporation (2002) 'Banana workers protest over possible imports' *Rural News*, 12 April. Online. www.abc.net.au/rural/news/stories/s529905.htm (accessed 27 September 2002).
Barrett, H.R., Ilbery, B.W., Browne, A.W. and Binns, T. (1999) 'Globalization and the changing networks of food supply: the importation of fresh horticultural produce from Kenya into the UK', *Transactions of the Institute of British Geographers*, 24: 159–74.
Biosecurity Australia (2002) *Importation of Fresh Bananas from the Philippines: Draft IRA Report*, Canberra: Department of Agriculture, Fisheries and Forestry.
Biosecurity Australia (2004) *Implications of Fresh Banana Imports from the Philippines: Revised Draft of Import Risk Assessment Report*, Canberra: Department of Agriculture, Fisheries and Forestry.

Bonlac Foods, Murray Goulburn Co-operative, Tatura Milk Industries and Warrnambool Cheese and Butter Factory (2003) *Comments on the Draft Import Risk Assessment: Importation of Bananas from the Philippines*, submission to Biosecurity Australia on Draft IRA Report, June 2002 (unpublished).

Borrell B., Ruby M. and Vincent D. (1993) *Inquiry into the Banana Marketing System in Australia*. Sydney: Horticultural Research and Development Corporation.

Brenner, N. and Theodore, N. (2002) 'Cities and the geographies of "actually existing neoliberalism" ', *Antipode*, 34 (3): 349–79.

Brimeyer, B.L. (2001) 'Bananas, beef and compliance in the World Trade Organization: the inability of the WTO dispute settlement process to achieve compliance from super-power nations', *Minnesota Journal of Global Trade*, 10: 133–71.

Bunt, C. (2002) 'Supply chain management in the Australian banana industry – a case study', *Acta Horticulturae*, 2: 433–35.

Burch, D. and Goss, J. (1999) 'Global sourcing and retail chains: shifting relationships of production in Australian agri-foods', *Rural Sociology*, 64: 334–50.

Chiquita Brands South Pacific Ltd (2003) 'Chairman's address to shareholders', 16 April.

Dicken, P., Kelly, P., Olds, K. and Yeung, H. (2001) 'Chains and networks, territories and scales: towards a relational framework for analysing the global economy', *Global Networks*, 1: 89–112.

Fagan, R.H. (2002) 'Bananas in chains? Globalisation, labour relations and the United States-European banana dispute', presented to Institute of Australian Geographers Conference, Canberra, July (available from the author: rfagan@els.mq.edu.au).

Food and Agriculture Organization (FAO) (2001) *Banana Information Note*. Rome: FAO, February.

Gereffi, G. (1994) 'The organization of buyer-driven global commodity chains: how US retailers shape overseas production networks'. In G. Gereffi and M. Korzeniewicz (eds) *Commodity Chains and Global Capitalism*, Westport (CT): Praeger, pp. 95–122.

Held, D., McGrew, A., Goldblatt, D. and Perraton, J. *et al.* (1999) *Global Transformations: Politics, Economics and Culture*. Cambridge: Polity Press.

Hermann, R., Kramb, M. and Monnich, C. (2003) 'The banana disputes: surveys and lessons', *Quarterly Journal of International Agriculture*, 42: 21–47.

Higgins, E. (2004) 'Why quarantine is a mean scene', *The Australian*, 19 February: 28.

Howitt, R. (2003) 'Contested concepts of scale in political geography'. In J. Agnew, D. Mitchell and J. Toal (eds) *A Companion to Political Geography*. London: Sage.

Hughes, A. (2000) 'Retailers, knowledges and changing commodity networks: the case of the cut flower trade', *Geoforum*, 31: 175–90.

Innisfail Advocate (2003) 'Banana ban ruling appeal irks farmers', 17 July: 3.

James, S. and Anderson, K. (1998) 'On the need for more economic assessment of quarantine policies', *Australian Journal of Agricultural and Resource Economics*, 42: 425–44.

Krinks, P. (2002) *The Economy of the Philippines: Elites, Inequalities and Economic Restructuring*. London: Routledge.

Leslie, D. and Reimer, S. (1999) 'Spatializing commodity chains', *Progress in Human Geography*, 23: 401–20.

McMahon, J.A. (1998) 'The EC banana regime, the WTO rulings and the ACP', *Journal of World Trade*, 32: 101–14.

O'Loughlin, T. (2002) 'Clean green machine bogged by rotten bananas', *Sydney Morning Herald*, 2 September: 4.

Ponte, S. (2002) 'The "latte revolution"? Regulation, markets and consumption in the global coffee chain', *World Development*, 30: 1099–122.

Pritchard, B. (1999) 'Australia as the supermarket to Asia: governments, territory and political economy in the Australian agri-food system', *Rural Sociology*, 64: 284–301.

van de Kasteele, A. (1998) 'The Banana Chain: the macroeconomics of the banana trade', paper presented at International Banana Conference, Brussels, May (unpublished).

16 Global cocoa sourcing patterns

Niels Fold

Introduction

Shifts in the global geography of cocoa sourcing patterns provide distinctive marks in the histories of key production areas. Explaining these shifts, however, has proved troublesome. On the one hand, 'economistic' explanations focus on rational behaviour by cocoa farmers. These approaches can be traced to Weymar (1968) who, in his classical contribution, explains these processes in terms of specific government incentives combined with farmers' rational behaviour – new plantings are influenced by the real price received by the cocoa producer, the real price of competitive crops and the real costs of new plantings – and long-term fluctuations determined along conventional cobweb logics (low prices, decline in planting, decline in production, supply deficit, increasing prices, new plantings, supply surplus, declining prices, etc.). This line of thinking remains widespread among multilateral organizations (see for instance International Trade Centre (ITC) 2001) despite its obvious limitations. Most notably, the fluctuating but steady growth of global cocoa production (by about one million tonnes) since the mid-1980s at the same time as prices have declined or stagnated (except for a short recovery period in the mid-1990s) indicates that a more complex suite of factors influence the dynamics of cocoa production, compared to what is assumed within rational 'economic man' models.

Consequently and on the other hand, some researchers have argued that environmental conditions in the cocoa frontier provide a more compelling explanatory set of factors for shifts in global cocoa sourcing. These arguments have been developed most comprehensively by the French economist François Ruf. His ideas are spread in numerous research reports and papers, and are elaborated upon most completely in the introductory chapter to his book *Cocoa Cycles* (Ruf 1995). There, he seeks to encapsulate the basic laws of cocoa supply, and to explain the shifts between cocoa production centres at the farm, regional, national and global level.[1] His basic starting point is that global cocoa supply is *not* determined by prices. What really matters for global cocoa supply is the existence of scarcely populated virgin forest areas that are relatively easy to clear and can be transformed into cocoa smallholdings or plantations. Planting cocoa on virgin forestland opens up the possibility to secure the vital 'forest rent' in the initial and first phases of cocoa cultivation. The concept of 'forest rent' conceptualizes the important advantages obtained by producers, in particular smallholders, by using land where the costs related to control of weed, soil fertility, soil moisture, pests, diseases and dry winds are very low compared to mature cocoa fields. It is the ability of producers in vacant virgin forest areas to operate at far lower costs than producers in mature areas that dictates the shifts between the major supply centres. The exploitation of the forest rent enables new areas to sustain and increase production in periods of falling prices, while production in high-cost (mature) areas gradually stagnates or decreases. In the long term, Ruf argues that this cycle explains the ebb and flow of production from one region to another.

Both the 'economistic' and 'environmentalist' approaches, however, fail to consider centrally the political economy of international trading companies and industrial processors. Noting this lack of regard for these dynamics, this chapter qualifies dominant understandings of global cocoa sourcing patterns by arguing that since the mid-1990s, the major global players in the global cocoa–chocolate value chain have increasingly determined the dynamics of the frontier. The large contract manufacturers of cocoa-based ingredients and the branded manufacturers of chocolate products increasingly have been involved in the organization of cocoa production on a world scale, in order to sustain an emergent global modular production network (Fold 2002; Sturgeon 2002). One reason for the increasing upstream involvement by the global giants is linked to potential supply barriers in the medium term. New and alternative growing areas are increasingly difficult to identify, and the previous reliance on frontier expansion has needed to be supplemented by conscious efforts to re-conquer degraded cocoa areas in order to maintain a (real or potential) surplus supply and, thereby, the present buyer-driven governance structure. These constraints need to be taken into account in order to understand the significance of recent initiatives to organize and coordinate chain-wide activities, including the appearance of a new, comprehensive form of private regulation on a global scale that incorporates commercially oriented non-government organizations (NGOs).

Behind the frontier: the persistence of cocoa farming smallholders

Cocoa is a prime example of a tropical commodity that is almost completely consumed in the North (i.e. in the industrialized countries of Europe and North America). Production and consumption of cocoa is also characterized by a high degree of concentration in the number of countries involved. On a global scale, the dominant importers of cocoa are the EU and the US. Country-wise, the US constitutes about 19 per cent (in volume terms) of global imports while the 'big seven' in the EU (Germany, the Netherlands,

France, UK, Belgium, Italy and Spain) make up 52 per cent. On the export-ing side, the Ivory Coast clearly dominates the picture with 48 per cent of total exports, followed by Ghana and Indonesia. Together, these three coun-tries account for 76 per cent of global exports. If Nigeria, Cameroon, Malaysia and Brazil are added, the cumulated share is close to 93 per cent of global exports.[2]

Most of the cocoa exported from the Ivory Coast and Ghana, and virtu-ally all of the cocoa exported from Nigeria and Cameroon, is destined for the EU.[3] In contrast to these African countries, Indonesia is completely dependent on the US market, and Brazil and Malaysia more or less had left the EU market for cocoa beans by the mid-1990s, Malaysia leaving a couple of years before Brazil. Hence, cocoa-bean supplies have become increasingly regionalized. South-East Asia is linked to the US market (Figure 16.1); Africa's cocoa exports are completely dominated by the EU (Figure 16.2), and since the mid-1990s, exports of cocoa from Latin America have declined, due primarily to the almost complete withdrawal of Brazil. Somewhat surprisingly, there is no clear market orientation for Latin American beans towards the North American market (Figure 16.3).

Parts of the explanation on the changing global cocoa supply patterns are found in the social and environmental dynamics of the frontier, as described by Ruf (1995). First, the model yields an insight into the general nature of potentially explosive contradictions between ethnic groups. Forest rents are

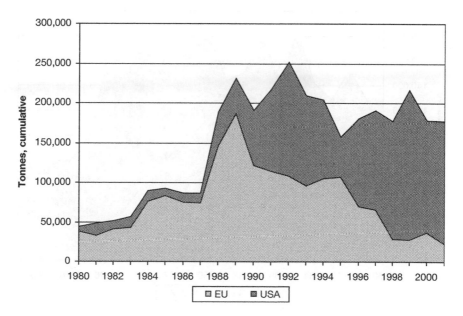

Figure 16.1 South-East Asian cocoa exports, 1980–2001, tonnes

Source: OECD ICTS International Trade by Commodity Database.

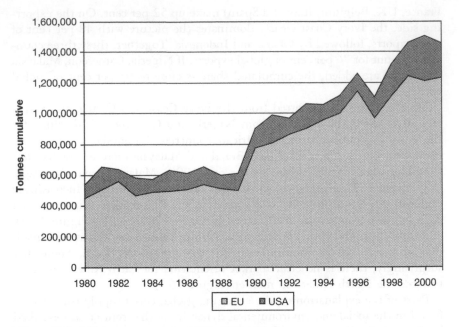

Figure 16.2 African cocoa exports, 1980–2001, tonnes

Source: OECD ICTS International Trade by Commodities Database.

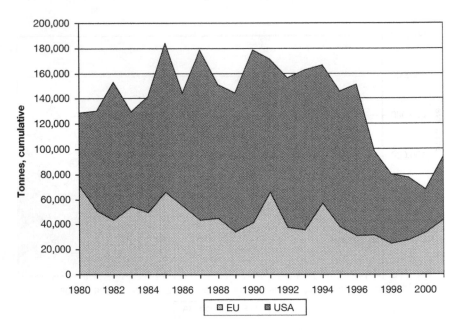

Figure 16.3 Latin American cocoa exports, 1980–2001, tonnes

Source: OECD ICTS International Trade by Commodity Database.

only exploited and frontiers only expand if surplus labour is available or can be mobilized through migration from another region or country. As the comparative work of Ruf and his colleagues show, all the major cocoa booms in the recent decades have materialized due to *migration waves*, sometimes sequenced in terms of different ethnic groups. Initially the migrants work for the indigenous population under some form of sharecropping arrangement or as wage-labourers, before acquiring their own land through various credit mechanisms. Alternatively, they may obtain virgin forestland in the first place by exchanging cash for land rights. However, over time these arrangements can lead to tensions along ethnic lines. As indigenous landowner households age, rural–urban migration of children can erode the supply of family labour, thus bringing into sharp focus questions of intergenerational land transfers. These tensions are observed in Sulawesi (Li 2001), the main region for Indonesian cocoa production, where highly entrepreneurial migrants have encroached on land belonging to indigenous hillside farmers. This type of conflict also seems to be a major component in the recent outbreak of civil war in the Ivory Coast just before the 2002–3 harvest season. Deeply embedded contradictions exist between indigenous groups and foreign migrants from countries in the Sahel (Burkina Faso, Mali). In the first phase of the migratory flow they were backed by a policy of 'the land belongs to those who cultivate it', formulated by the first government that took over after independence (Chauveau 1995). During the 1990s, rivalry for scarce land resources increased, leading to attempts to 'renegotiate' earlier land arrangements and sometimes even violent clashes between the ethnic groups in the countryside. The conflict was further intensified in the late 1990s when the rural land tenure law was reformed, in order to recognize customary law. Increasingly, the conflict has been incorporated into a complicated political–military struggle between new alignments of political parties, with potentially destructive consequences for economic and political stability in the country (Crook 2001; Woods 2003).

Second, Ruf's framework also emphasizes the devastating effects of *pest and diseases* in aging cocoa-growing regions with pronounced mono-cropping practices. The virtual disappearance of Brazilian cocoa beans in global exports is a salient case in point. Since the late 1980s, the 'witches' broom' disease has wiped out nearly three-quarters of the production in Bahia, previously responsible for about 85 per cent of the total annual crop. Many cocoa farms in Bahia are commercially managed plantation-like farms depending on the high input of capital and labour. The collapse of the cocoa sector has resulted in widespread poverty among approximately 90,000 farm labourers who have lost their jobs, and catapulted the regional economy into a severe recession (Bright 2001). Comprehensive replanting programmes with 'witches' broom'-resistant cultivars were started in the late 1990s, but it is highly questionable whether a renewed Brazilian cocoa sector will be economically viable in the future due to the relatively high wage level.

Similarly, the relatively high *labour costs* in prolonged periods of low and stagnating world market prices have been the main cause behind the disappearance of Malaysian beans from the global scene. Cocoa is primarily cultivated in Sabah, one of the two Malaysian states in northern Borneo, and most of cocoa cultivation takes place on plantations owned and managed by private companies (Jarrige 1995). The private sector responded to the unprecedented high prices in the late 1970s and early 1980s with massive increases in plantings, but a decade or so later they cleared the cocoa trees and replanted the land with oil palms (Lee and Musa 1999). In addition to lower labour costs, palm oil offered far brighter prospects for future profits, and oil palm areas in Sabah and Sarawak (the other Malaysian state on Borneo) were expanded rapidly during the 1990s (Sutton 2001; Leigh 2001).

Evidently, low global *cocoa prices* have played an important role in these outcomes. Most commercial plantations closed down their cocoa activities during the 1990s. At prevailing price levels, the scope for commercial plantations is extremely limited by relatively high labour costs and high risks linked to pest and disease attacks, which easily spread in vast, mono-cropped areas. It is even questionable whether a significant price increase (such as the one in 2002) will constitute a sufficient incentive for capitalist agriculture to return into cocoa production. The medium- to long-term prospects need to be more promising: a 2004 forecast anticipates a return to a situation where supplies exceed grinding demands after a short-lived supply squeeze in 2002 (ED & F Man 2004).

Low global prices impact differentially on smallholders, who, because of family labour, do not necessarily have to bear wage labour costs. Moreover, currency devaluations in producer countries cushioned the consequences of declining world market prices for smallholders, because farmers may actually experience an increase in nominal producer prices even in periods with sustained price decline. In all the major producing countries in West Africa, local currencies have been devalued as part of structural adjustment packages. The Indonesian currency was also devalued in connection with the so-called Asian crisis in the late 1990s. In such a situation, farmers may increase production and thereby further stimulate price declines on the world market due to fundamental global supply and demand mechanisms. Moreover, it is likely that devaluations influenced the downward spiral of world market prices, as international traders negotiating in US dollars (or other currencies with international convertibility) are able to factor in the new exchange rate in their offers. For instance, during the late 1990s when global supply and stock declined, world market prices did not increase, indicating the role of local currency devaluations. There is a considerable time lag before the devaluation works its way through the economy and results in increasing inflation, thus eroding producer gains and decreasing production.

As a corollary, global supply of cocoa is heavily based on smallholders – African smallholders in particular – producing in a context of potentially disruptive social and ethnic conflicts. At the same time it has to be realized that the cocoa frontier is coming close to an absolute spatial barrier in the sense that except for a few countries (Vietnam, Papua New Guinea) potential cocoa-growing areas are hard to identify.

The continued importance of independent smallholders in this traditional global agri-commodity chain contrasts with recent observations on the marginalization and exclusion of smallholders in cultivation and exports of fresh horticultural products from developing countries (see, for instance, Dolan and Humphrey 2000; Barrett *et al.* 2004). Due to the implementation of strict food safety and labour standards set by supermarkets (and public institutions) in the Northern countries, large commercial farms increasingly dominate the production in the Southern end of these chains (see also Barling and Lang, this volume).

Beyond the frontier: changing structures and dominant actors in the global value chain

During the recent decade, profound concentration and centralization processes have taken place in the cocoa–chocolate value chain, resulting in a dominant position for a handful of international grinding companies and international chocolate manufacturers (Fold 2002). Some of the chocolate manufacturers are giant corporations in the global food industry, specializing in branding and marketing a number of different consumer products, including chocolate (Nestlé, Kraft), while others are specialized in chocolate-based products (Mars, Hershey's, Cadbury, Ferrero). These companies also have in-house grinding capacity in order to maintain the ability to manufacture intermediate proprietary chocolate products. A similar division of company types is found among the international grinders where some (Barry Callebaut, Blommer) are specialized in basic cocoa products (variations of paste, powder, butter as well as generic and customized chocolate products) while others are versatile agri-food companies, where cocoa processing is just one line of business among other agri-processing activities (ADM, Cargill). The latter group of companies is able to transfer and adapt technical, organizational and managerial competences from one business line to another. Important changes in logistics have occurred in the global chain due to the introduction of high-volume, bulk transportation by chartered ships and 'flat' storage of beans in warehouses in importing countries. These practices are much more cost-efficient than previous systems of storage and liner transportation of beans in jute bags.

As indicated in Figure 16.4, the main actors in the global cocoa–chocolate value chain are not vertically integrated. Chocolate manufacturers increasingly have outsourced the production of intermediate cocoa products, while grinders have sold off chocolate manufacturing divisions of the companies they have acquired over the years. A relatively new phenomenon, however, is the trend towards backward integration of the dominant grinders into exporting operations, often in the form of direct control if not majority ownership of local exporting companies. Until now, none of the grinders have gone into domestic trading operations (i.e. purchasing beans directly from producers). Instead, a hierarchy of local traders – some of which are being financed on a more-or-less daily basis by the grinders – carry out these activities. A few of the specialized warehouses from the European 'cocoa-hub' in Amsterdam

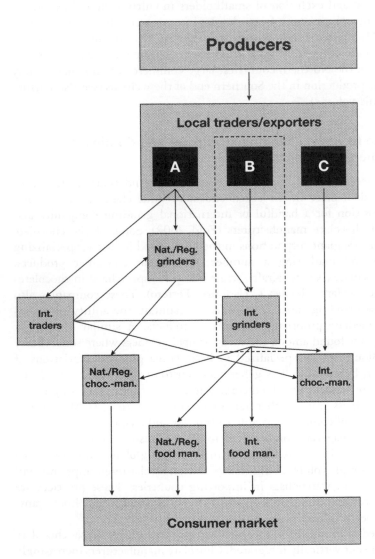

Figure 16.4 Structure and actors in the global cocoa–chocolate value chain

Source: Author.

Note: Int. = International; Nat./Reg. = National/Regional; Choc. = Chocolate; Man. = Manufacturers.

have established facilities in West Africa from where they service the international grinders. Most of these operations previously were carried out by marketing boards and licensed companies in the African cocoa-producing countries, but as a consequence of structural-adjustment-related liberalization, most of these parastatals have been dismantled in all countries except Ghana (Fold 2002; Losch 2002).

One of the reasons for increasing upstream engagement by grinders is related to quality concerns. Cocoa beans from African countries used to have a significant premium compared to beans from Brazil and South-East Asia. The difference was caused by the (generally) more careful after-harvest treatment of beans by African farmers (i.e. farmers allow the beans sufficient time for fermentation and drying under appropriate conditions). In contrast, post-harvest labour input on plantations or medium-scale farms in Latin America and South-East Asia are far lower due to higher labour costs; beans are not given sufficient time for fermentation, and the drying process is carried out with different kinds of machinery or open fire instead of sun-drying.

With the liberalization of the marketing boards in African producer countries, existing institutions for purchase and quality control have vanished. Instead, private traders now strive to increase the rate of capital turnover in order to maximize profits. As a result, local traders are willing to buy beans of dubious quality if they are considered saleable. Hence, farmers sell insufficiently fermented and dried beans to willing buyers and the beans have to be dried in the harbours before they are shipped (Varangis and Schreiber 2001). The earlier premium on African beans has therefore decreased and origin differentials are reduced; there is a trend towards a 'global bean' of inferior quality.

Some observers argue that this trend is a simple reflection of a liberalized environment in which market participants determine the quality/price relationship and minimum quality standards are obeyed. Industrial customers are simply not willing to pay the premium for high-quality beans as this attribute has been rendered dispensable due to: (1) development of process (grinding) technology and (2) new transport practices where all kind of beans are mixed (Gilbert 1997; Gilbert and Tollens 2002).

In contrast, the Association of the Chocolate, Biscuit & Confectionery Industries of the EU (CAOBISCO) is trying to change the current standard commercial contracts for international trade in cocoa beans (CAOBISCO 2002). According to the association, current contracts do not take into account that (1) beans are now of inferior quality due new farmer practices and (2) beans with different qualities are subsequently mixed in the ports to reach acceptable limits under existing contracts. Taken together, the result is shipments of beans of widely different quality and size. Hence, there is a need for new contracts listing key quality criteria (covering off-flavours, moisture, homogeneity, fat quality and packing material) and the methods for determining whether these criteria are met.

The different views and concerns over quality and the nature of customer demand is related significantly to the different actors in the global cocoa–chocolate value chain and the variations in technological capacity for the manufacturing of intermediate goods. The dominating international grinders are more able to handle beans of inferior quality because their plants are highly flexible and composed of advanced equipment. In contrast, chocolate manufacturers (particularly in Europe) who still process parts of their intermediate chocolate products have not invested heavily in new equipment but carry out operations on relatively obsolete plants. They source beans by themselves or buy from international trading companies in addition to supplies from the independent grinders (see Figure 16.4). Hence, lower bean quality is obviously a significant problem for the smooth operation of their cocoa processing plants.

Penetrating the frontier: towards global private regulation of cocoa sourcing?

New regulatory developments in the global cocoa–chocolate value chain are evolving in response to an ongoing erosion of two forms of *public* regulation. First, various types of state regulation – mostly in the form of marketing boards – have been dismantled as part of structural adjustment programmes. Irrespective of the former inefficiency of various state activities (including extension services, input supplies, pest and disease management, quality control etc.), these have now largely disappeared as few private companies have replaced the state institutions. The regulatory linkages between agriculture and the state are diminishing, leaving the sector open for free enterprise and potential manipulation for short-term profit.

Second, global public regulative institutions are also losing importance. In 2001, the Sixth International Cocoa Agreement was agreed upon by a number of cocoa exporting and importing countries. Compared to previous international agreements, the new agreement appeals primarily to the goodwill of member countries and other interested parties while the (in principle) important regulatory mechanisms (e.g. export quotas, buffer stocks) have been completely removed (United Nations Conference on Trade and Development (UNCTAD) 2001). Although there may have been valid reasons for their abolition,[4] the subsequent arrangements basically remove the last substantial vestiges of public regulation at the global level.

Into these voids, distinctive forms of *private* regulation have evolved. Importantly, these initiatives are more than mere image cultivation, despite their extensive usage of rhetoric and development buzzwords. Contemporary supply barriers, structural changes and quality concerns define the nature of these new forms of regulation designed and undertaken by the major actors in the global cocoa–chocolate value chain.

Pre-eminently, the maintenance of smallholder involvement is absolutely vital in order to secure stable and abundant supplies of cocoa of adequate quality to the global industry. Even if new cocoa frontiers are opened, additional production and exports will not be of a sufficient volume to replace potential losses if existing cocoa-growing regions are allowed to degenerate further. Therefore, all of the major companies and business associations are increasingly cooperating in order to 'fill up what has been hollowed out', i.e. to revive cocoa production among smallholders in (over) mature cocoa areas, primarily in West Africa but also in Indonesia. The organizational structure is still somewhat opaque and best conceptualized as being in an early and innovative phase.

With regard to the recent evolution of private global regulation, a decisive turning point seems to be the turmoil raised by press reports in 2001 about the use of child labour in West African cocoa sectors under conditions that violated internationally accepted standards. In the US, pressure from policymakers was mounting and consumer organizations voiced their concern to chocolate manufacturers (Fold 2004). As a response, the major transnational cocoa processing and chocolate manufacturing companies considered common ways to take action. The first result was the formal establishment of an international protocol in late 2001 committing European and North American industry associations and major individual companies to ensure that by July 2005, cocoa is grown 'without abusive child or forced labour'. In addition, industry representatives together with a number of international NGOs and labour unions signed a joint statement that outlined a timetable with a number of important milestones. These include the implementation of an independent survey of labour conditions in the major West African cocoa producing countries and the setting up of a joint foundation to promote responsible cocoa farming and eliminate abusive labour practices by the implementation of a certification system by July 2005.[5]

In July 2002, the joint foundation was established under the title of the 'International Cocoa Initiative'. This seems to be the basic organizational structure that is going to coordinate and monitor the projects and support programmes initiated by the signatories of the international protocol. In November that same year, a region-wide programme to promote responsible cocoa labour practices was launched. The programme is based on the concept of 'pilot programmes', and activities can be grouped into two categories. The first one is designed and managed by the International Labour Organization (ILO). It focuses on responsible labour practices and child labour interventions by protecting at-risk youth, training local authorities, disseminating knowledge and raising awareness about labour issues. The second category addresses what is called 'the critical underlying issues' such as the health and vibrancy of local cocoa farming communities, including individual farmers' needs for assistance. These programmes are managed through a network already established in relation to the implementation of the Sustainable Tree Crop Program (STCP), which is managed by the International Institute of Tropical Agriculture (IITA), a UK-based NGO with a regional office in Cameroon. Components include the establishment and training (finance, marketing) of farmer groups, the improvement of cultivation practices, the dissemination of technology for pest and disease management, and education of farmers via radio.

A pilot programme for education of 'master' trainers has already been successfully implemented and a curriculum for the training of trainers has been developed (see IITA 2003). It is the intention that these 'master' trainers then train a group of selected farmers who then become trainers. Each trainer is responsible for the establishment of a 'field school' with 25–30 farmers in a village community, initially being assisted by experienced personnel from the national extension services. If this pyramidal principle really works as expected, it may be possible to implement the certification scheme before the tight deadline of July 2005. It remains to be seen whether this remarkable progress in programme implementation is sustainable, or whether it will confront problems that are revealed at a later stage. Suffice to say that the issue of labour standards has been incorporated in ongoing efforts to organize African smallholders in a (post-liberalization) private regulatory structure of potentially unprecedented scope.

A parallel but more modest programme is operating in Sulawesi, and another one has been established recently in Brazil. Both of these programmes are under the auspices of the World Cocoa Foundation, which, despite its name, is an association consisting of a limited number of the large companies with activities in the US cocoa–chocolate industry. The association was established in the late 1990s and has coordinated a number of different research programmes ranging across breeding and germplasm maintenance, integrated crop management, and so-called regional farmer support programmes ('SUCCESS Alliances') in South-East Asia (Indonesia, Vietnam, and the Philippines) and Latin America.[6] These latter programs are implemented primarily by two US-based NGOs; Agricultural Cooperative Development International (an entity formed by US cooperatives and farm credit banks), and the Volunteer in Overseas Cooperative Assistance, an NGO originally specializing in volunteer assistance to developing countries.

A key issue for further research on similar forms of global private regulation is to understand the nature, role and relative strength of the NGOs that are involved in these kinds of initiatives. The variations among NGOs are extremely wide and increasingly include the offspring from trade associations and networks of business interests, in addition to the traditional humanitarian and religious organizations. More knowledge on the constituencies, strategies and influence of participating NGOs is warranted by the fact that they seem to be crucial for the establishment of links between global industry actors and aid agencies. One way or another, the lion's share of the above-mentioned programmes at farm or community level is ultimately financed by public institutions.

Another key issue is how and why major actors in particular global value chains organize themselves in this kind of global, chain-wide structure. After all, many of the companies in the global cocoa–chocolate industry have a

long history of intense competition for global dominance. For instance, anec-
dotal evidence suggests that ADM and Cargill compete 'cut-throat' whenever
and wherever possible, while Brenner (1999) has reported the numerous
clashes between Mars and Hershey's. One explanation is the seriousness of
the supply situation in the medium term as it has been outlined above.
Another explanation, perhaps a necessary condition, is that over the years
the global cocoa–chocolate industry has been involved in other 'common
battles', for instance on nutritional issues, cocoa butter substitutes and the
recent child labour issues. Perhaps this previous experience of mutual
exchange of concerns, coordination of statements and positions and lobbying of public institutions for common interests have resulted in some
kind of shared 'cooperative capital' that is not found in other global value
chains. Moreover, it seems probable that the consolidated structure of the
global cocoa–chocolate chain is of decisive importance to the ability to act
in common.

Conclusion

The restructuring of global cocoa sourcing patterns over the past decade
incorporates some striking contradictions. On the one hand, an almost
annually recurrent surplus of cocoa on the world market caused prices to be
depressed during the 1990s. More recently, modest reductions of cocoa
production in the 2001–2 harvest year and recent dramatic military events
in the Ivory Coast have resulted in a price hike not observed in almost 20
years. This phenomenon suggests a latent set of global cocoa-supply problems, brought into the open by the worldwide closing down of cocoa
plantations, the scope of attacks by pests and diseases in mature cocoa regions
and the escalation of socio-economic problems and ethnic clashes in major
cocoa-producing areas. For cocoa smallholders, this short-lived price rise
provides a much-needed respite from difficult economic conditions and an
incentive to maintain production, notwithstanding the fact that the almost
continuous devaluations of producer countries' local currencies over the past
decade has cushioned the cocoa processors, chocolate manufacturers and
consumers in the North from drastic price hikes.

Hence at the turn of the century, the industry started being much more
concerned about supply conditions, in addition to price. Various initiatives
spanning all processing segments in the chain have commenced, primarily
involving the US-based companies. A main objective is to secure a steady
flow of cocoa from smallholders in the tropics, in a context of fading public
regulatory mechanisms, both at the international and national levels. A recent
incident in relation to reports on violation of international labour standards
resulted in severe pressure for common action. The global industry responded
via a remarkable organizational innovation, designed and set up in a very
short course of time. Labour issues are now integrated with 'traditional' extension activities in order to secure fair labour conditions, to protect production

and income in cocoa communities, to manage pest and diseases and (not the least important) to introduce a certification scheme by 2005. These initiatives are perhaps the early phase of a hitherto unseen incorporation of cocoa smallholders in contract farming schemes. It is a fascinating example of wide-ranging private regulation encompassing the supply roots of a global chain, implemented in what seems to be a swift, efficient and bold move that underlines the corporate sector's self-image of 'no-nonsense' engagement – so far.

Notes

1   The model is based on abstractions over frontier trajectories in all the major cocoa-producing countries, many of which Ruf has an intimate knowledge through his empirical work over the last couple of decades. For reasons of space it is not possible to include all the complexity and different aspects of the model in this chapter. Furthermore, the model is quite opaque as it consists of 15 main components, including mutually influencing 'cycles' (of labour, smallholders, cocoa trees etc.), different – sometimes overlapping – concepts of 'rents' and various state policies (levies, demographic regulation) and marketing systems. Apparently, Ruf's ambition was to continue with the work towards the construction of a formal quantitative model, hence the occasional listing of rather strange 'equations'.

2   The figures are three-year averages (1997–9) and include trade in both raw and processed beans, the latter having been converted into raw beans (see UNCTAD (2001) for details). Basically, processing of cocoa consists of grinding the beans into cocoa paste and subsequently pressing the paste into cocoa butter and cocoa cake; the cake is usually ground into cocoa powder before it is used in the food industry. In the manufacturing of chocolate (the major final use of cocoa), cocoa paste is mixed with cocoa butter and sugar, sometimes adding milk and other ingredients (nuts, fruit, etc.).

3   The following paragraph is based on data from the OECD's database 'International Trade by Commodities'.

4   These arrangements turned out to be useless in practice because of free-riding, bad financial management and outright resistance among member countries (Gilbert 1996).

5   For details see, for instance, the homepage of World Cocoa Foundation (www.chocolateandcocoa.org).

6   Other activities include an on-line store selling cocoa video commercials, ballpoint pens, T-shirts, mugs – and 'Sid the chocolate bear' ('Squeeze his hand and hear him say: I looove chocolate . . .').

References

Association of the Chocolate, Biscuit & Confectionery Industries of the EU (CAOBISCO) (2002) *Cocoa Quality*. Réf: 166–58sj-Rev.5 (Mimeo). Brussels: CAOBISCO.

Barrett, H.R., Browne, A.W and Ilbery, B.W. (2004) 'From farm to supermarket. The trade in fresh horticultural produce from sub-Saharan Africa to the United Kingdom'. In A. Hughes and S. Reimer (eds) *Geographies of Commodity Chains*, London: Routledge, pp. 19–38.

Brenner, J.G. (1999) *The Emperors of Chocolate: Inside the Secret World of Hershey and Mars.* New York: Broadway Books.

Bright, C. (2001) 'Chocolate could bring the forest back', *World Watch Magazine*, November/December: 17–28.

Chauveau, J.P. (1995) 'Land pressure, farm household life cycle and economic crisis in a cocoa-farming village (Cote d'Ivoire)'. In F. Ruf and P.S. Siswoputranto (eds) *Cocoa Cycles: The Economics of Cocoa Supply*. Cambridge: Woodhead, pp. 107–23.

Crook, R.C. (2001) 'Cocoa booms, the legalization of land relations and politics in Cote d'Ivoire and Ghana: explaining farmers' responses', *IDS Bulletin*, 32: 35–45.

Dolan, C. and Humphrey, J. (2000) 'Governance and trade in fresh vegetables: the impact of UK supermarkets on the African horticulture industry', *Journal of Development Studies*, 37 (2): 147–76.

ED & F Man (2004) *Cocoa Market Report*, No. 231, 30 March.

Fold, N. (2002) 'Lead firms and competition in 'bi-polar' commodity chains: grinders and branders in the global cocoa–chocolate industry', *Journal of Agrarian Change*, 2 (2): 228–47.

Fold, N. (2004) 'Spilling the beans on a tough nut: liberalisation and local supply system changes in Ghana's cocoa and shea chains'. In A. Hughes and S. Reimer (eds) *Geographies of Commodity Chains*. London: Routledge, pp. 63–80.

Gilbert, C.L. (1996) 'International commodity agreements: an obituary notice', *World Development*, 24 (1): 1–19.

Gilbert, C.L. (1997) *Cocoa Market Liberalization: Its Effects on Quality, Futures Trading and Prices*. London: The Cocoa Association of London.

Gilbert, C.L. and Tollens, E. (2002) 'Does market liberalization jeopardize export quality? Cameroonian cocoa, 1995–2000', *Discussion Paper No. 3224*, Centre for Economic Policy Research.

International Institute of Tropical Agriculture (IITA) (2003) *STCP Newsletter* (June).

International Trade Centre (ITC) (2001) *Cocoa: A Guide to Trade Practices*. Geneva: UNCTAD/WTO.

Jarrige, F. (1995) 'Ivorian and Malaysian cocoa supply: a comparative study of structures and performance'. In F. Ruf and P.S. Siswoputranto (eds) *Cocoa Cycles: The Economics of Cocoa Supply*. Cambridge: Woodhead, pp. 249–79.

Lee, M.T. and Musa, M.J. (1999) 'Future prospects of cocoa production in Malaysia', *The Manufacturing Confectioner*, December: 90–5.

Leigh, M. (2001) 'The new realities for Sarawak'. In C. Barlow (ed.) *Modern Malaysia in the Global Economy*. Cheltenham: Edward Elgar, pp. 119–32.

Li, T.M. (2001) 'Planting trees and losing ground: the cocoa boom and land transfers in Sulawesi', *Paper for the Euroseas Conference*, 6–8 September.

Losch, B. (2002) 'Global restructuring and liberalization: Cote d'Ivoire and the end of the international cocoa market?', *Journal of Agrarian Change*, 2 (2): 206–27.

OECD ICTS International Trade by Commodity Database. Online: http://new. sourceoecd.org.

Ruf, F. (1995) 'From "forest rent" to "tree capital": basic "laws" of cocoa supply'. In F. Ruf and P.S. Siswoputranto (eds) *Cocoa Cycles: The Economics of Cocoa Supply*. Cambridge: Woodhead, pp. 1–53.

Sturgeon, T. (2002) 'Modular production networks: a new American model of industrial organisation', *Industrial and Corporate Change*, 11 (3): 451–99.

Sutton, K. (2001) 'Agribusiness on a grand scale – FELDA's Sahabat Complex in East Malaysia', *Singapore Journal of Tropical Geography*, 22 (1): 90–105.

United Nations Conference on Trade and Development (UNCTAD) (2001) *International Cocoa Agreement, 2001*, TD/COCOA.9/7. Geneva: UNCTAD.

Varangis, P. and Schreiber, G. (2001) 'Cocoa market reforms in West Africa'. In T. Akiyama, J. Baffes, D.F. Larson and P. Varangis (eds) *Commodity Market Reforms. Lessons of Two Decades*, Washington (DC): The World Bank, pp. 35–82.

Weymar, F.H. (1968) *The Dynamics of the World Cocoa Market*, Cambridge: MIT Press.

Woods, D. (2003) 'The tragedy of the cocoa pod: rent-seeking, land and ethnic conflict in Ivory Coast', *Journal of Modern African Studies*, 41 (4): 641–55.

17 The world steer revisited

Australian cattle production and the Pacific Basin beef complex*

Bill Pritchard

Introduction

Narratives of international beef sector restructuring hold influential sway within recent research on agri-food globalization. In the mid-1980s, Steven Sanderson persuasively coined the phrase 'the world steer' to describe a trajectory of restructuring in which large-scale Fordist-style meat production systems were developed in Third World destinations to service affluent Northern markets (Sanderson 1986). In Sanderson's vision, 'the world steer' paralleled the much vaunted 'world car'. He proposed that international restructuring of the beef sector had entered 'a new phase, qualitatively different from previous modes of external influence' in which: 'The international economic integration of the nineteenth century, which relied primarily on commodity circulation, has been supplanted by a holistic integration of the cattle sector in production' (Sanderson 1986: 124).

Central to these processes was the control of international beef supply chains by agri-food transnationals with capacities for the sourcing of product from multiple destinations. In Latin America, from whence Sanderson drew his empirical data and inspiration, progression towards these outcomes was seen as a historical transformation in the continent's agri-industrial development, with detrimental implications for national economic development and food security.[1]

In the 1990s, the kernel of Sanderson's arguments provided inspiration for research into the restructuring of the beef sector in the Pacific Basin. The focus on integration within the Pacific Basin, as opposed to global integration, acknowledges the global geo-economics of the beef industry. Disease barriers, trade agreements and transport costs have contributed to the existence of two relatively separate trade-production circuits in the international beef industry: a Pacific circuit and an Atlantic–European circuit. Reciting Sanderson's model, Ufkes (1993: 219) interprets the 1988 US–Japan Beef

* The research for this chapter was funded through Australian Research Council Discovery Grant 'The Spatial Construction of Food Commodity Chains'. Thanks to Pinar Cabadag for research and other assistance.

Liberalization Agreement and the Australia–Japan Beef Liberalization Agreement as an important stage in a progressive de-nationalization of production systems in which: 'Circuits of transnational agro-food capital now integrate regions within core, peripheral and semi-peripheral countries into highly complex food commodity chains.' As evidence for this, she points to how these agreements were pre-empted by significant Japanese investment in the US beef sector, and by US and Japanese investment in the Australian beef industry. For Ufkes, these investments forged a highly integrated cross-Pacific beef complex (Ufkes 1993: 226–7). A similar conclusion is reached by Jussaume (1996) in research that documents the cross-Pacific investments of Japanese beef processing and trading firms. In the post-liberalization era of the Pacific Basin beef complex, he contends that firms, rather than nations, have become the central institutional actors for managing and coordinating trade (1996: 71). Finally, further developing these themes several years later, Francis (2000: 531) seeks to explain 'the conversion of national beef industries within the Pacific Basin into a *geographically coherent industry*' (italics mine), asserting that 'a global beefpacking industry in the Pacific Basin has emerged, [although] national markets for beef persist'.

Taken together, Sanderson's seminal paper and later published research give the suggestion of a uni-directional and pervasive set of transformations towards the trans-Pacific integration of beef production and trade, consistent with an historical reconfiguration of the international conditions for profitability in the sector. According to this general line of argument, international mergers and acquisitions in this industry have led the Pacific Basin beef sector to become dominated by transnational firms, and international trade relations have been transformed to become a set of intra-firm transactions dependent upon the execution of multiple sourcing strategies by these same firms.

It is both timely and important to revisit this scholarship. The research by Ufkes (1993) and Jussaume (1996) was prompted by a series of major international mergers and acquisitions that took place in the late 1980s and early 1990s. Francis's (2000) research does not significantly update these events, despite its later publication. The objective is not so much to use the benefit of hindsight to point out shortcomings of that earlier scholarship but to use the passage of time to document more recent processes of restructuring and consolidation in this sector. Specifically, this chapter asks whether beef production in the Pacific Basin has been oriented increasingly towards satisfying import demand by Japanese (and to a lesser degree, other North-East Asian) markets, via trade relations saturated by transnational corporate coordination and control?

To bring evidence to this task, attention is given to the restructuring dynamics connecting Australian beef production with the Japanese market. In the Pacific Basin beef sector, Australia is positioned as a low-cost supplier. Extensive access to rangelands has enabled Australia to become the world's largest beef exporter, without recourse to production subsidies of the type that characterizes Northern agriculture. Therefore, consideration of internationalization processes in the Australian beef sector represents a geo-economic variant of the themes and issues analysed elsewhere in this book.

Transformations in the Pacific Basin beef complex during the 1980s and early 1990s

The structural foundations of the Pacific Basin beef complex underwent important transformations during the late 1980s and early 1990s as Japan, which had previously imposed severe restrictions on beef imports, opened its market significantly. The origins of this process were caught up within the international trade politics of the Uruguay Round of the General Agreement on Tariffs and Trade (GATT). American interests sought to open the Japanese market for beef as part of a broader agenda to reduce the widening trade deficit between the US and Japan. These efforts culminated in the 1988 liberalization agreements mentioned above, under which the Japanese government increased import quotas and lowered tariffs (Ufkes 1993: 222–6).

The significance of these agreements is readily apparent. In 1975, Japan imported only 85,000 tonnes of beef, representing just 15.6 per cent of Japanese beef consumption (FAO 2003). During the next twelve years Japan continued to implement a highly restrictive beef import regime, despite intense pressure from the key export nations of the US and Australia. Japanese beef imports grew by just 10,750 tonnes per year over this period, so that by 1987 they had inched upwards to only 214,000 tonnes (FAO 2003).

Implementation of the 1988 agreements triggered an explosive transformation to these arrangements. In the three years that followed their signing, import quotas were relaxed sufficiently to allow a further 180,000 tonnes of beef to enter Japan, representing an increase in annual import volumes of 83 per cent. Thenceforward, the Japanese government committed itself to replacing absolute quota restrictions with a tariff-quota regime mandated by a schedule of tariff cuts. These policy changes effected a significant increase in Japanese beef consumption during the first half of the 1990s. The country's traditional reliance on seafood as a source of protein began to give way to red meat. From 1990 to 1995, Japanese domestic beef consumption increased by 37.2 per cent, from 1.055 million tonnes to 1.447 million tonnes, and the ratio of imports in total consumption grew from 48 per cent to 58 per cent (FAO 2003).

In terms of the Pacific Basin beef complex, the important point to be made about these developments is that they executed a transformation to the structural composition of the industry. As documented and retold by Ufkes (1993), Jussaume (1996) and Francis (2000), Japanese liberalization was accompanied by considerable offshore investment by Japanese interests in American and Australian beef-packing firms, and a secondary flow of US investment in the Australian beef-packing sector. In effect, Japanese beef and trading interests sought to pre-empt the impacts of liberalization by gaining control of the offshore production that would become increasingly important to the nation's procurement system. Interpreting these developments, Ufkes presages their significance: 'New forms of agribusiness control of regional agricultures have emerged with consequences for new structures of international beef trade and for the international division of labour in the beef commodity chain' (1993: 226).

Four firms spearheaded the Japanese investment in the Australian beef sector. In 1988, Nippon Meat Packers Ltd, Japan's largest beef company, purchased a half-share in Thomas Borthwick and Sons Ltd, at the time, Australia's fifth largest beef-processor. In 1990, it acquired full control of the processor (Asahi News Service 1990). In 1989, the Japanese general trading company C. Itoh purchased a 40-per-cent interest in R.J. Gilbertson Ltd, Australia's third largest beef-packing company. Following corporate manoeuvres in Japan, this investment was later held by Sumikin Bussan, a smaller, specialist trading firm. Also in the late 1980s, the Mitsubishi group of companies, which had operated a feedlot business in Australia since 1970, purchased Mid-Coast Meats, Australia's eighteenth largest beef-processor. And a few years later, in 1995, the Japanese trading houses, Mitsui and Zenchiku purchased a 40-per-cent stake in the G. & K. O'Connor meat-works, the sixteenth largest meat processor in Australia at the time.

Close in pursuit of the initial investments by Japanese interests, two of Australia's iconic beef companies were acquired by other foreign interests. In 1992, the US transnational food corporation ConAgra Inc. purchased a 50.1 per cent stake in Australian Meat Holdings Pty Ltd (AMH). This company was (and remains) Australia's largest beef processor, and at the time was owned by the Australian conglomerate Elders IXL Ltd. A failed management buy-out of Elders IXL by its chief executive led to creditors selling the company's assets, including the AMH business.[2] In 1994, ConAgra purchased the remaining equity in this business to attain total control. Also in 1994, the Chinese International Trust and Investment Corporation (CITIC), a Chinese state-owned enterprise, purchased Metro Meats. This firm was owned previously by the Adelaide Steamship Company Ltd, another conglomerate that collapsed in the early 1990s. Through these events, ownership structures in the Australian beef-processing industry were radically transformed in a short space of time during the late 1980s and early 1990s.

There is little doubt that the rapid entry of these transnational interests in the Australian beef sector, accompanied by the expansion of feedlot production systems to service the rapidly expanding Japanese market, represented a profound shift in the direction of this sector. Although the presence of foreign investment was not wholly new in this sector (dating from the colonial period, British interests had extensive investments in the Australian beef industry), these acquisitions appeared to suggest a new phase of the industry, in which local production systems would be integrated more deeply within international trade networks. Coming at a time when deliberations of the Uruguay Round of the GATT appeared to hold out the possibility of ushering in a neo-liberal regime of global agriculture, it is hardly surprising that contemporary researchers interpreted these developments in ways that portended an historical juncture in the sector. So to re-state the question posed earlier, to what extent has this trajectory unfolded? In answering this, two foci will be attended. First, have North-East Asian markets exercised a continued 'pull' on the industry, so that domestic production systems have become oriented increasingly to servicing these markets? And second, has transnational capital increasingly saturated the production and trade networks of the Australian beef system?

North-East Asian markets for beef

Researchers writing in the early and middle 1990s tended to interpret the 1988 Japanese beef liberalization agreements as the first stage in a progressive *de-nationalization* of beef production and consumption spaces in the Pacific Basin. It was assumed that Japan's demand for beef would grow at a rapid rate, as its markets were opened and as diets were transformed. Indeed, the title of Francis's (2000) article – 'eating more beef' – explicitly positions these assumed developments as the focal point of the restructuring dynamics of the Pacific Basin beef complex.

Importantly, however, these expectations have not come to pass. The second half of the 1990s witnessed a dramatic terminus to the growth phase of beef consumption in Japan. As illustrated in Figure 17.1, the growth of beef and veal import volumes slowed considerably in the middle 1990s. Combined with lower prices for beef and the depreciation of the yen, this trend contributed to a significant fall in the value of Japan's beef imports after 1995 (Figure 17.2). In 2001, the US dollar value of Japan's beef imports was approximately 30 per cent lower than its level in 1995. Whereas total Japanese beef consumption grew by 37.2 per cent between 1990 and 1995, it remained virtually static in the following half-decade (Food and Agricultural Organization of the United Nations (FAO) 2003). Moreover, as also indicated in Figures 17.1 and 17.2, other regional markets did not provide consistent and sustained growth markets for beef exports. The growth phase of the South Korean market peaked in the early 1990s and remained relatively static afterwards, excepting a single-year surge in beef imports in 2000. Hong Kong and China remain relatively minor import markets.

These outcomes are intimately connected to international trade politics in the second half of the 1990s. Under the 1988 beef liberalization agreements, Japan replaced quotas in 1991 with a 70-per-cent tariff. By the late 1990s, in accordance with Uruguay Round commitments, this had been reduced to 38.5 per cent (Meat and Livestock Australia Ltd 2001: 8). To be sure, this represented considerable liberalization compared with previous arrangements. Nonetheless, these tariffs remain a significant restriction upon imports. Given

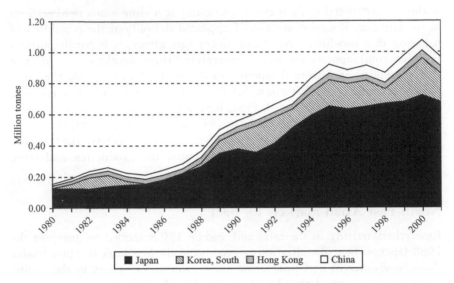

Figure 17.1 The volume of beef and veal imports to North-East Asian countries, 1980–2001

Source: FAO 2003.

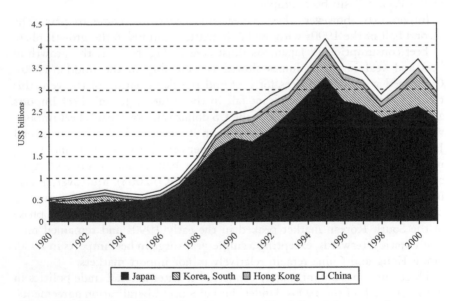

Figure 17.2 The value of beef and veal imports to North-East Asian countries, 1980–2001

Source: FAO 2003.

the intransigency of current multilateral negotiations on agriculture and the unwillingness of the Japanese government to unilaterally reduce tariffs further, this rate has become a semi-permanent fixture in the landscape of the Japanese beef sector. Its overall effect has been to dampen the further incursion of beef imports into the Japan.

In any case, it seems apparent that the Japanese market for beef has matured and therefore offers relatively limited opportunities for further growth. Japan's population is growing slowly and ageing rapidly, and its economy has been depressed since the early 1990s. Moreover, the nation's embrace of Western foods appears to be waning. In December 2002, for example, the McDonald's chain announced the closure of 176 restaurants in Japan, in the context of a 2.3 billion yen (US$19.1 million) annual loss (Reuters News Service 2002). In the first few years of the twenty-first century, food scares further contributed to the sombre outlook for beef consumption in Japan. In 2000 Japan experienced its first outbreak of foot-and-mouth disease for almost a century, and in 2001 Japanese authorities confirmed cases of bovine spongiform encephalopathy (BSE, or 'mad cow disease') in its domestic herd. Then in 2002, in the midst of the BSE crisis, Nippon Meat Packers Ltd was prosecuted for mislabelling imported beef as domestic product so that it could fraudulently receive monies under a government buy-back scheme. These events contributed to a collapse in consumer confidence for domestically produced beef, and led to imports taking a larger share of a smaller market. Facing this environment, in July 2003 the Japanese government controversially utilized WTO Safeguard provisions and announced an increase in beef tariffs from 38.5 per cent to 50 per cent.[3]

The persistence of tariff protection for the Japanese beef sector, in the context of a market that appears to offer relatively limited further potential for growth, brings into focus the changed trajectory of the Pacific Basin beef complex. North-East Asia, and Japan in particular, has not provided the propulsion that earlier researchers assumed it would. This environment of weaker growth in beef demand and the maintenance of significant protection for the domestic beef sector heavily qualifies the contemporary application and appropriateness of the 'world steer' model to explain restructuring trends in the Pacific Basin beef complex.

Recent ownership changes in the Australian beef-processing sector

The changed conditions for beef exports to North-East Asia since the mid-1990s affected the industry structure within the Australian beef-processing sector. In 1995, following the entry of Japanese, American and Chinese investment in the sector, foreign interests controlled 46.5 per cent of the production volume of the 25 largest meat-processing firms (Table 17.1).[4] This represented a considerable share of industry output and, in the opinion of the research scholarship cited at the outset of this chapter, foretold an industry structure that would be saturated by transnational corporate control and coordination. However, subsequent developments have complicated these prognostications. In 2001, foreign interests accounted for 41.7 per cent of the production volume of the top 25 processors, a *lesser* share of total industry tonnage than they did in 1995.[5] To understand the reasons for this reversal, a fine-grained perspective on industry change is required.

Changes in ownership patterns within the Australian beef-processing sector since the mid-1990s represent the expression of three developments. First, a number of privately owned Australian firms have expanded aggressively (Table 17.1). These include Teys Brothers (Holdings) Pty Ltd (owned by the Teys family group), Bindaree Beef, Midfield Meats and the Consolidated Meat Group (owned by the Packer family, whose patriarch is Australia's richest individual). Reflecting their successes, in 2002 Teys and CMG merged their businesses into a single entity, owned privately by the two parties. This has become Australia's second-largest beef processor, and represents a substantial reorganization of market power towards domestic interests.

Second, since the early 1990s, the industry has proven increasingly un-attractive for relatively smaller foreign investors whose participation has been dependent upon rates of return considerations. These tendencies have been illustrated to best effect through the exit of CITIC and Sumikin Bussan, two foreign investors that were attracted to the industry on the basis of expec-tations that buoyant export growth conditions of the early 1990s would continue. CITIC's participation in the industry lasted only two years. Expectations that it could build an export beef business from Metro Meats were unfounded in the generally difficult business climate of the middle 1990s. In 1996 and 1998, CITIC closed and then sold its 'Metro Meats'

Table 17.1 Ownership of the 25 largest meat processors in Australia, by tonnage, various years from 1995 to 2001

	1995	1999	2000	2001
Australian-owned	53.5	59.0	55.5	58.3
of which:				
Australian private-owned	n.p.	46.0	48.2	53.0
Australian public-owned	n.p.	8.0	2.6	2.3
Producer cooperatives	n.p.	5.0	4.7	3.0
Foreign-owned	46.5	41.0	44.5	41.8
of which:				
US-owned	16.5	23.0	25.2	25.4
Japanese-owned	14.0	16.0	14.6	11.5
Other	16.0	2.0	4.7	4.9
Total	100.0	100.0	100.0	100.1

Source: ProAnd Associates (2003).

Note: n.p. = data not published separately. Percentages in the year 2001 do not add up to 100.0 because of rounding.

business. Also during this period, Sumikin Bussan increased and then divested its interests in the Australian beef sector. In 1996 it extended its 40-per-cent stake in R.J. Gilbertson to 100 per cent and renamed the business SBA Pty Ltd. But in 1999, the company closed the most important of its three processing facilities (in Altona North, Victoria) and in 2002 the entire business was sold back to Australian interests, which renamed the entity the Tasman Group.

Third, related to the processes described above, foreign participation in the industry has largely contracted to three corporate groups: Nippon Meat Packers, AMH and Cargill (Table 17.2). The operations of the two largest of these firms (Nippon Meat Packers and AMH), in particular, underline the role of *transnationality* as a strategic source of competitive advantage that explains their on-going participation in this industry.[6] This is seen with clarity in the financial performance of Nippon Meat Packers during this period (Table 17.3). From 1994 to 2000, revenue earned from the company's Australian operations trended downwards and profit rates tended to be very low. On the basis of these data alone, it may be difficult to imagine how Nippon Meat Packers could justify its continuation in the Australian industry. However, interpretations of these data need to be conditioned by a wider appreciation of the firm's international operations. Nippon Meat Packers' Australian activities represent only the first stage of an integrated series of intra-firm transactions that link Australian beef production to Japanese supermarkets. Beef exported out of Australia is sold to the Japan Food Corporation, a Japan-based affiliate of the company's parent (Nippon Meat Packers 2001: 28). Consequently, low or negative profitability in Nippon Meat Packer's Australian beef-processing operations may be an unimportant issue for the corporate group as a whole, if these production arrangements contribute to profitability elsewhere in the production chain.[7] The recent experience of AMH also intimates the importance of perceiving Australian operations in a wider, transnational context. AMH is the largest beef processing company in Australia by a considerable margin, and exports the vast majority of its output. Moreover, like Nippon Meat Packers, the company has reported a relatively low profit rate over recent years.[8] Yet, during this period AMH fulfilled a particular role within the broader international strategy of its corporate parent, ConAgra. AMH was a supplier of low-grade beef to the US market, an important resource for ConAgra's domestic operations in America. Furthermore, ownership of AMH allowed ConAgra to source the Japanese market from either Australia or the US (Australia: Department of Foreign Affairs and Trade 2001: 145). This provides the type of intra-firm multiple sourcing opportunities envisaged in Francis's conceptualization of a 'geographically coherent industry' in the Pacific Basin.

These developments – the expansion of domestic interests, exit of smaller foreign firms and the continuing participation and expansion of larger corporations with transnational, intra-firm production-trade networks –

Table 17.2 The six largest beef processors in Australia, 2001 (AUS$ millions)

	Ownership	Domestic sales	Export sales	Total sales
Australian Meat Holdings	US-owned (ConAgra)	330	1,870	2,200
Nippon Meat Packers	Japanese-owned	121	689	810
Teys Brothers	Australian (private)	58	522	580
Consolidated Meat Group	Australian (private)	120	425	545
Bindaree Beef	Australian (private)	180	270	450
Cargill Foods Australia	US-owned (Cargill)	150	150	300
Total of top six		959	3,926	4,885

Source: Feedback (2002).
Note: Financial data for Consolidated Meat Group is an estimate, based on industry averages.

Table 17.3 Financial performance of Nippon Meat Packers Australia Pty Ltd, 1994–2002

Year	Column A – receipts from customers before income tax (AUS$'000)	Column B – operating profit (%)	Profit ratio (B/A) (%)
1994	633,663	10,790	1.70
1995	513,494	–9,366	–1.82
1996	491,113	14,609	2.97
1997	372,599	–3,004	–0.81
1998	428,221	930	0.22
1999	498,788	1,555	0.31
2000	577,518	6,036	1.05
2001	689,847	28,311	4.10
2002	816,010	36,727	4.50

Sources: Nippon Meat Packers Australia Pty Ltd (various years).

signify a generally more complex set of restructuring processes than pre-supposed by scholarship written in the immediate aftermath of the 1988 Japanese beef import liberalization agreements. Whereas the basic thrust of these researchers' arguments remains valid (that is, a considerable proportion of the Australian industry has been integrated within transnational corporate networks oriented towards the export of beef to markets in the Pacific), the nuances of restructuring since the mid-1990s reflect a more recent trade and production architecture. Compared to earlier expectations, there has been a stalling in: (i) the growth of North-East Asian demand for beef; (ii) processes of de-nationalization in these domestic beef markets; and (iii) the proportion of the Australian industry incorporated within transnational corporate networks. Attention now turns to the question of how these empirical trends should be interpreted.

The contemporary condition of the Pacific Basin beef complex

Australian beef production is connected to North-East Asian markets via an entanglement of domestic and transnational interests. In the late 1980s and early 1990s, the allure of rapid growth in Japanese beef imports attracted transnational interests to the Australian beef sector, which proceeded to incorporate a considerable proportion of the industry within trans-Pacific, intra-firm networks. These players continue to exercise major influence over the industry, but the stillborn character of Japanese beef liberalization has cast limits on further incursions by transnational corporate interests.

Central to these developments is the fact that beef processing is a generally low-profit industry characterized by considerable risk and uncertainty. Beef processing is styled on Fordist production techniques where improvements to profitability are achieved most readily through the reorganization of production into larger-scale facilities that exploit economies of scale. In North America, where these developments can be observed with greater clarity, there has been a wholesale relocation of processing capacity to plants designed in ways to maximize processing throughput. In conjunction with these developments, the workforce in this industry has been deskilled, casualized and paid less (Stull and Broadway 2004). At the same time, moreover, the international beef sector operates in accordance with climatic and market cycles, which generate risk and uncertainty for owners of processing capacity.

These contexts tend to dictate the terms under which transnational capital is attracted to the industry. The recent history of the beef sector tends to suggest that transnational corporate interests exercise greatest influence as an agent of restructuring when the capture of profits from rapid market growth requires the acquisition and control of processing capacity. In conditions of more modest growth, there are fewer incentives to tie up shareholder equity in the direct ownership of processing facilities that, in general, offer a relatively low rate of return.

In the beef sector, these impulses have been seen most dramatically in ConAgra's 2002 decision to divest its beef division. In a deal valued at US$1.4 billion, ConAgra sold 54 per cent of its beef-processing operations (including AMH) to Hicks, Muse, Tate and Furst, a Dallas-based leveraged buyout fund. Through this divestiture (and a subsequent sale of the company's poultry business), ConAgra sought to transform itself from a diversified agrifood corporation to a specialist in value-added, branded foods. These processes have broader parallels in the food industry, where during the past decade an increasingly wide bifurcation of corporate strategies has emerged between the ownership of branded products and intangible capital on the one hand and the scale-production of processed foods on the other (Pritchard 2000). Research by Pritchard and Burch (2003) on the processing tomato industry, for example, reveals a division between transnational companies such as H.J. Heinz Co. and Unilever, which focus on the marketing of branded foods, and national-centred companies, most commonly owned through private and/or family interests, which focus on the large-scale production of tomato paste as a standardized and globally traded food ingredient.

For the substantive issues of this chapter, these developments bear sensitive interpretation. Evidently, the evolution of ownership arrangements in the Australian beef-processing sector speaks to a pluralistic set of processes involving different forms of capital and competition. Slower growth of the Japanese market and relatively difficult conditions for profitability has tended to discourage the overt and direct integration of the Australian beef-processing sector into transnational corporate networks. Instead, these processes have been mediated by new investment models based around privately held corporate structures, linked to international markets through a range of organisational arrangements.[9] These structures do not unambiguously reflect a 'holistic integration of the cattle sector in production' (to chime with Sanderson) or a 'geographically coherent industry' (as suggested by Francis) but suggest a set of capitalist processes that are much more selective in the ways that agri-food production systems are incorporated within the logic of globalization.

Conclusion

This chapter has sought to reveal important qualifications to the Pacific Basin beef complex as described in earlier research. Pre-eminently, it underlines how international trade politics have provided a defining historical condition for the contemporary patterning of agri-food production and trade. Viewed with the benefit of hindsight, it is apparent that the scholarship cited at the outset of this chapter focused too intently on the institutional capacities of transnational corporations as agents of production and trade coordination, and thereby encouraged an inflated perception of the process of de-nationalization. In contrast, the global-scale politics of trade relations have intervened in such a way as to limit processes of transnational integration within the Pacific Basin beef complex, and specifically, in Australia. Whereas the construction of internationally coordinated production arrangements provided the key propulsive dynamic in this sector during the late 1980s and early 1990s, this has not been the case more recently. As such, the deployment of Sanderson's 'world steer' model needs updating, so that it is sensitized to the contemporary geo-economics and geo-politics of agri-food production and trade, if it is assist our understanding of this sector.

Seen in its widest context, this chapter takes its inspiration from the contention that the neo-liberal dream of open markets for agriculture and food will not become a reality. The distinction this chapter makes relates to the way that research on this industry in the early 1990s conflated *tendencies* with *structures*. In the early 1990s, researchers identified and documented a set of emerging and important processes connected with Japan's rapidly growing appetite for beef imports and the ways that transnational corporations were responding to these developments through strategies of offshore investment. However, seen from the vantage point of the early 2000s, these processes were a historically contingent expression of the search for profit by global agri-food corporations. In the more recent elaboration of the international politics of food, the Australian beef sector is no longer a magnet for investment by transnational agri-food interests.

These international political realities need to inform scholarship on cross-continental agri-food systems. The concept of 'the world steer' provides an understanding of the political–economic composition of the global beef sector if the neo-liberal dream of unfettered markets was taken to its logical conclusion (cf. Le Heron this volume). Tendencies towards these outcomes are invoked in contemporary agri-food restructuring, but the model itself does not represent an accurate portrayal of the current situation in this sector. The current era requires historically sensitized and empirically contemporary accounts of global agri-food complexes, if agri-food scholarship is to reflect accurately the economic and political impulses of the age.

Notes

1   Sanderson's argument is that these international trade relations constructed national dependencies on the production and import of feed grains, in exchange for the export of beef in a crowded and volatile international marketplace. He cites statistical evidence suggesting an inverse relationship between the expansion of the export beef sector and the nutritional exigencies of the rural poor.

2   Elders IXL also owned the Foster's brewing enterprise. Its chief executive officer at the time was Mr John Elliot. Elliot was later indicted with a range of criminal charges relating to his tenure at the helm of Elders IXL, but these were not proven in court.

3   The WTO Safeguard provisions allow member countries to temporarily implement higher tariff rates to curb increased imports. Under the WTO rules, tariffs can be increased so long as import growth exceeds 17 per cent in a given three-month period. In this instance, Japan's beef imports grew by 34 per cent between April and June, 2003 (Agra Europe 2003).

4   These data on the 25 largest beef processing firms provide the only reputable estimation of ownership share in the industry. They are collected annually by the consultancy firm ProAnd Associates for publication in the industry journal *Feedback*. It needs to be recognized that (i) data are based on volume levels, not the total sales or profit levels in the industry (hence, giving a bias towards bulk processors of relatively lower-valued cuts); (ii) they include all red meat production, and (iii) these data relate only to the 25 largest firms. Although proportions vary annually, the 25 largest firms generally account for approximately 80 per cent of total Australian beef production. Assuming that smaller firms not included in the 'top 25' ranking tend to be mainly Australian-owned, these data therefore over-state the proportion of foreign ownership in the industry as a whole. I would like to thank ProAnd Associates for making available some previously unpublished components of these data.

5   Francis (2000: 546) wrongly states that the majority of the Australian beef processing industry is foreign-owned. Moreover and remarkably, he makes no mention of the evolving ownership structures in the industry during the 1990s, despite the fact that

his research was published many years after the initial ConAgra and Nippon Meats acquisitions and that data was readily accessible on this issue.

6   Some mention, in passing, needs to be made of Cargill. This firm diversified into the Australian beef industry in the late 1980s following a long-standing presence in the Australian grain-trading sector. Its operations remained relatively small until an acquisition in 1998. Compared to other major beef companies Cargill's operations are oriented more greatly to the domestic market (see Table 17.2).

7   During the past decade, the formulation of these intra-firm arrangements has given rise to accusations aired in the media that Nippon Meat Packers may be engaging in transfer pricing, that is, keeping the prices of Australian beef exports artificially low in order to register low taxable income in Australia (for example, see Dickie 1996; Jolly 1997). In the absence of definitive evidence on this matter it is difficult to evaluate these issues. Published financial statements during this period do not indicate the payment of corporate income tax during the second half of the 1990s, and the 2000 report notes that the company was the subject of an audit by the Australian Taxation Office. In the subsequent two years, Australian revenue and profitability jumped markedly, and the company paid significant sums of corporate income tax in Australia.

8   The ratio of earnings (before abnormals and income tax) to revenue was 3.7 per cent in 2001, and 1.84 per cent in 2002. Source: AMH Pty Ltd documents lodged with the Australian Securities and Investments Commission (018347984, 017700247).

9   It might be surmised that privately held corporate structures may accommodate the difficulties of unpredictability and risk better than those with common stock equity, because private owners tend to have greater ability to subsume immediate rate of return considerations to longer term strategic imperatives.

References

Agra Europe (2003) 'Japan imposes emergency tariff on pork imports too', *Agra Europe*, 2065 (1 August): M/8.

Asahi News Service (1990) 'Japanese firm acquires fourth Australian meat processor', *Asahi News Service Report*, 27 April.

Australia: Department of Foreign Affairs and Trade (2001) *Agrifood Multinational Corporations in Asia*. Canberra: DFAT.

Dickie, P. (1996) 'Why farmers are dirt poor', *Sunday Mail* (Brisbane), 27 October: 75.

Feedback (2002) 'New companies join Feedback's Top 25 ranking', *Feedback*, 3 (9): 8–11.

Food and Agricultural Organization of the United Nations (FAO) (2003) 'Food trade statistics'. Online: www.faostats.org (accessed 30 September 2003).

Francis, R. (2000) 'Eating more beef: market structure and firm behaviour in the Pacific Basin beefpacking industry', *World Development*, 28 (3): 531–50.

Jolly, J. (1997) 'Japanese investment reflects confidence in commercial future', *The Weekend Australian*, 18 January: 30.

Jussaume, R. (1996) 'Agricultural trade, firms and the State: extrapolations from the case of Japanese beef imports', *International Journal of the Sociology of Agriculture and Food*, 5: 66–84.

Meat and Livestock Australia Ltd (2001) *Global Beef Liberalisation: Magellan Project*. Sydney: Meat and Livestock Australia.

Nippon Meat Packers (2001) *Annual Report*. Nippon Meat Packers: Tokyo.

Nippon Meat Packers (various years) Annual financial statements lodged with the Australian Securities and Investments Commission (documents 018416552, 017820417, 009900808, 017820415, 017820414).

Pritchard, B. (2000) 'The tangible and intangible spaces of agro-food capital', unpublished paper presented at the 10th International Rural Sociological Association World Congress, Rio de Janeiro, Brazil. Available from the author (School of Geosciences, University of Sydney NSW 2006 Australia).

Pritchard, B. and Burch, D. (2003) *Agri-food Globalization in Perspective: International Restructuring in the Processing Tomato Industry*. Aldershot: Ashgate.

ProAnd Associates (2003) Personal communication, 12 September.

Reuters News Service (2002) 'Japan: McDonald's expects a loss', *New York Times*, 21 December: 3.

Sanderson, S. (1986) 'The emergence of the world steer: internationalisation and foreign domination in Latin American cattle production'. In F.L. Tullis and W.L. Hollist (eds) *Food, The State and International Political Economy*. Lincoln (NB): University of Nebraska Press, pp. 123–48.

Stull, D.D. and Broadway, M.J. (2004) *Slaughterhouse Blues: The Meat and Poultry Industry in North America*. Belmont (CA): Wadsworth.

Ufkes, F. (1993) 'Trade liberalisation, agro-food politics and the globalisation of agriculture', *Political Geography*, 12 (3): 215–31.

Index

Page numbers in *italic* type indicate tables or figures.